Pupil.	Form	Date Issued
P. SMITH	VI L.	9/10/72
K. Hill	VI Lower	6-9-76

The United States
and Canada

HUTCHINSON ADVANCED GEOGRAPHIES

The United States and Canada
W. R. Mead and E. H. Brown

The Monsoon Lands of Asia
R. R. Rawson

Africa
A. B. Mountjoy and C. Embleton
with a contribution by W. B. Morgan

IN PREPARATION
The British Isles
E. M. Yates

Australia and New Zealand
R. L. Heathcote and R. G. Ward

The United States and Canada

A Geographical Study of Regional Problems

W. R. Mead, D.Sc. (Econ.), Ph.D.
and
E. H. Brown, M.Sc., Ph.D.
Professors of Geography, University College London

HUTCHINSON EDUCATIONAL

HUTCHINSON EDUCATIONAL LTD
178–202 Great Portland Street, London W1

London Melbourne Sydney Auckland
Johannesburg Cape Town
and agencies throughout the world

First published 1962
Second impression December 1963
Third impression September 1964
Fourth impression April 1967
Fifth impression (revised) October 1970

Printed in Great Britain by litho on smooth wove paper
by Anchor Press, and bound by Wm. Brendon,
both of Tiptree, Essex

ISBN 0 09 103750 6 *(net)*
0 09 103751 4 *(non-net)*

Contents

THE NORTH AMERICAN CONTINENT

Plates

Maps and Diagrams

Preface

It is fashionable for texts on North America to be the product of joint authorship and to offer resulting puzzles in author detection. It is not proposed to spoil the reader's fun by providing a solution for this book. Suffice to say that both authors began their North American experiences in Canada under similar wartime circumstances: one in Ontario, the other in Alberta and Saskatchewan. Both have been in North America several times since, including a journey together on the occasion of the XVIIth Congress of the International Geographical Union in Washington D.C. in 1952. In 1953 the one struck deeper roots as a visiting professor to the University of Minnesota and in 1954 the other followed suit at Indiana University. For both, the geographical appreciation of North America provides a foil to more serious research in other areas.

Consequently, this book makes no claim either to comprehensiveness or to profundity: it offers a personal appreciation of some of the subcontinent's problems in their regional context. The maps and diagrams have been drawn by John Bryant of University College London, who found work on them sufficient to stimulate him to emigrate to the University of Victoria, B.C. Three considerations have guided their selection. First, a series of systematic distribution maps is introduced to complement the essentially regional approach. Secondly, it is assumed that the reader will have access to a good atlas, but that he is unlikely to have topographical maps to hand. It has therefore been felt better to include a range of examples of different types of landscape and settlement based upon American and Canadian topographical maps of fairly large scale. Thirdly, a transect diagram has been constructed for a part of each of the major regional subdivisions. A key showing the location of transects, maps and plates is included as Figure 69.

For permission to use material from copyright sources for maps and diagrams the authors are greatly indebted to the following: George Allen & Unwin Ltd., Figure 27; the American Geographical Society, Figures 33 and 38; British Columbia Natural Resources Conference and the Department of Mines and Resources, Figure 63; the Department of Mines and ·Technical Surveys, Ottawa—Geographical Branch—Figures

54, 60A and B, and 66; *Economic Geography*, Figure 12; the Geographical Press, a division of C. S. Hammond & Co., Maplewood N.J., Figures 3, 6 and 24; Harper & Brothers, Figure 4; McGraw-Hill Book Co., Inc., Figure 9; and Rand McNally & Co., Figure 20.

Among halftones, emphasis is placed on vertical air photographs because they play an increasingly important role in the interpretation of North American landscapes and in the planning of their use. Especial thanks for supplying photographs are extended to Dr. Norman Nicholson and to Dr. F. Marschner. Dr. Nicholson, as head of the Geographical Branch of the Department of Mines and Technical Surveys in Ottawa, has supplied Plates IX-XVI with the permission of the Royal Canadian Air Force. Dr. Marschner's photographs (III-IV), provided through the U.S. Department of Agriculture, were first printed in U.S.D.A. Publication No. 153, *Agriculture and Land Use in the U.S.A.*, a book indispensable to all students of American geography. Historical 'gracenotes' have been obtained through Dr. Herman Friis of the National Archive in Washington (Plate II) and the Champlain Society of Canada (Plate Ia).

We are indebted to Kenneth Wass of University College London, for additions to and amendments of the maps and diagrams.

If this volume communicates a little of the enthusiasm which its authors feel for the U.S.A. and Canada, if it helps a little towards the understanding of an unnaturally misunderstood part of the world, if it encourages a handful to read a little more deeply into the rich treasury of North American geographical literature, and if it stimulates a few to explore the unknown byways of a challenging land, it will have served its purpose.

W.R.M.
E.H.B.

February 11, 1962

TO THE MEMORY OF

Captain Alexander Maconochie, R.N., K.H. (1787–1860)

first Professor of Geography in Britain at University College London, who, like the authors, first made acquaintance with North America during wartime, he in the attacks on Washington and New Orleans in the war of 1812, they on more friendly visits in 1941.

Introductory

I

The Problem of an Approach

This book is a study of regional problems, but the biggest problem of all has been to apportion a hundred thousand words among them. The model cannot be that of John Banvard's *Panorama of the Mississippi* (1846), a three-mile-long canvas deficient in artistry if accurate in geography. It must be nearer to that of Thornton Wilder's play *The Skin of our Teeth*, sub-titled 'A History of Mankind in a Comic Strip'. Something of the cartoonist's method, with its distortion and omission, over-simplification and understatement, is perforce the result. The geographer is perhaps at his best when following the technique of Jane Austen as a miniaturist in ivory: he is apt to be at his worst when uttering 'the Big Bow-wow strain' of Sir Walter Scott. The broad sweep of the pen encourages subjective assertion and for better or worse emphasizes the study of geography as an art. It is only in the detailed investigation of cause and effect that geography can be an exact science.

A text such as this cannot pursue enquiry into detail; it can at best summarize enquiries by others. Many aspects of North American geography have been precisely investigated and a rich literature of findings has been published by learned societies. Behind investigation lies measurement, and in many respects measurement is central to geography as a science. North America offers facts and figures in abundance and the records of the land and its people are noted in the two following chapters. In the body of the text, however, statistics have generally been subordinated to personal observation.

North America invites observation—be it from skyscraper or vista dome. There is, however, too much to observe. The student of North America is immediately confronted by the problem of size and the related problem of distance. Admittedly the average American and Canadian, or innocent abroad in their lands, speedily disregards distance. Faced with a giant, he must take to seven-league boots. As St. John Crèvecœur wrote in his *Letters from an American Farmer* (1759), 'A European when he first arrives, seems limited in his intentions as well as in his views; but he very suddenly alters his scales.' There is no ultimate solution to this problem of scale. In the final instance the image of North America presented by any writer and the impression left upon the reader must be personal. The image will be an assembly of many fragments of truth, or of truth in process of change, for the picture changes not only with the viewer but with the time of viewing.

To observe this happening in the past is to understand that it is happening in the present. Two examples will serve to show how the climate of the age as well as the environment of the author influence the interpretation of an area. Reference may first be made to the 'colonial' phase of under-assessment, when the natural philosophers of Europe cast a somewhat jaundiced eye over Atlantic America and statesmen held the sugar islands of Caribbea in higher esteem than the empire of the St. Lawrence. The naturalist Georges Buffon believed that in the New World 'everything languishes, corrupts and proves abortive'. Nature, diminishing New World quadrupeds, also caused degeneracy in European bipeds who settled in America. The Abbé Raynal, in his *Philosophical History* (1779), felt that in British North America 'the spirit dries up like the body'; while Pehr Kalm in 1748–50 observed an arrested development. In contrast, there was the age of over-assessment associated with the nineteenth-century nationalism. William Gilpin, first governor of Colorado, had a golden vision of the Mississippi Valley. He visualized it as an 'expanded concave bowl, to receive and fuse into harmony whatsoever enters within its rim'. How much more fertile this mountain-rimmed garden of the New World than the mountain-rimmed 'salt-water waste' of the Old World Mediterranean. Inspired by Alexander von Humboldt, Gilpin identified an 'isothermal zodiac of nations' and it was a manifestation of destiny that the westward course of empire should carry leadership to 'the democratic republican empire of America'. Arnold Guyot, a Swiss immigrant contemporary of Gilpin who lectured on *The Earth and Man* (New York, 1887) in 1849, held a similar point of view:

> The fertility of [America's] soil, its position in the midst of the oceans between the two extremes of Europe and Asia, facilitating commerce with these two worlds: the proximity of the rich tropical countries of Central and South America, towards which, as by a natural descent, it is borne by the waters of the majestic Mississippi ... Science may attempt to comprehend the purposes of God as to the destinies of nations, by examining with care the theatre seemingly

arranged by Him . . . the order of nature is a foreshadowing of that which is to be.

Though they showed individual variations, North American geographers in the nineteenth century and after inclined strongly towards determinism. They range from the Ratzelian disciple Ellen Churchill Semple, as continually readable as she is perennially provocative, to N. S. Shaler, the Harvard geologist, with his vision of men growing to larger stature in north-west Washington and south-east Appalachia. By the time of Guyot and Gilpin the U.S.A. had achieved a political unity from coast to coast and had defined its landward boundaries. It then became aware of divisions within. The Civil War, behind which lay geographical as well as political differences, was the most violent expression of this division. Canada has suffered no civil war, but 'secession' is a word which has not been absent from its domestic vocabulary. The outward unity of the sub-continent is therefore shot through with inner differences which are rooted in the nature of the countryside and in the evolution of its peoples.

Geographies of North America, no less than studies in other disciplines of the sub-continent, have passed through a succession of phases. Systematists and regionalists have held successive sway, with the increasing refinement of research techniques urging forward changes in the approach to the subject. Whereas the geographers of a century ago saw in North America an inherent physical unity, those of today see a vast and varied terrain upon which has been imposed a striking human unity. This unity is inseparable from technical considerations. No part of the world has been more inventive than North America. No people have overcome the obstacles and distances which separate them from each other by more effective mechanical means. At the same time they have imparted a curious similarity to inherently dissimilar areas by spreading a veneer of common innovations over landscapes different in physical character. This fact impresses the consequences of living in an extensive area without restraint upon commodity movement, and within which, for political ends, common standards are promoted.

The nature of North America is the subject of an immense literature. Not surprisingly, it has given birth to a perceptive school of geographers —from pioneers such as Jedidiah Morse (1761–1826), through the 'versatile Vermonter' George Perkins Marsh (1801–1882) to contemporary humanists such as Carl Sauer and philosophers such as Richard Hartshorne. While North America has always had leaders in the study of geography, it was never more a hearth of ideas than at the present time. Chicago and Evanston, Washington State and Pennsylvania State are names with which to conjure among progressive geographers. It is in their institutions that quantitative studies have been pursued with the greatest vigour, that regional scientists have conceived their most advanced models and that the language of geography has been correspondingly elaborated. In many respects, the United States of America is the seat of a new geography. The new geography may have its shortcomings. Yet whatever they may be, there are many who might legitimately claim that the paradigm of geography in the second half of the twentieth century is the gift of America. It is not the purpose of this book to debate these issues. They simply provide an additional reason why geographers at large should be reasonably familiar with the land, the people and the problems of the North American sub-continent.

SELECTED READING

Ideas about North America as a land in which to live are pieced together from the literature of the time by

> A. Mumford Jones, 'The Colonial Impulse', *Proceedings of the American Philosophical Society*, 90, 2 (1946), 131–61
>
> G. Chinard, 'Eighteenth-century Theories of America as a Human Habitat', ibid, 91, 1 (1947), 27–57

An excellent anthology of earlier landscape appreciation is

> John Bakeless, *The Eyes of Discovery* (New York, 1950)

It is beautifully complemented by

> David Lowenthal, *The American Scene* (*G.R.* 58 (1968) 61–68).

Four contrasting domestic interpretations are those of

> Jedidiah Morse, *The American Universal Geography* (Boston, 1805)
>
> William Gilpin, *Mission of the American People* (Independence, Missouri, 1873)
>
> E. C. Semple, *American History and its Geographical Condition* (London, 1903)
>
> D. Lowenthal, *George Perkins Marsh, Versatile Vermonter* (New York, 1959)

Two gatherings of professional geographers have been guided through the U.S.A. under the aegis of the *International Geographical Union*. The official handbooks of 1911 and 1952 are worth studying, while a more informal journey is described in

> G. R. Stewart, *U.S. 40, Cross Section of the U.S.A.* (Cambridge, 1953)

Finally, two useful surveys are

> P. E. James and C. F. Jones, *American Geography, Inventory and Prospect* (Syracuse, 1950)
>
> Saul B. Cohen (ed.) *Problems and Trends in American Geography* (N Y 1967)

2

A Land of Many Characters

*The Record of the Land—Geometry in North American Geography
—The Rocks and their Arrangement—The Heights of Land and the
Ways of the Waters—A Survey of the Morphological Regions—The
Record of the Climate—Orography and Movements of Air—Oceans
and Climatic Depressions—Climatic Hazards—Types of Climate—
A Survey of Soils—The Nature of the Vegetation.*

North America is a land of many characters. They are expressions of its
diverse land forms, of its varied climates and many weathers, of its resulting
range of soils and consequent variety in vegetation. The natural elements
in consort give distinctive personalities to different places. The personality
of the land is fundamental to this regional text and the record of its
features is a prerequisite to its understanding.

THE RECORD OF THE LAND

Many different authorities have been responsible for measuring, defining
and analysing the physical background of North America. With the
dawn of the nineteenth century the U.S.A. initiated an era of planned
exploration. It was conducted principally from the office of the Adjutant
General, with the Secretary of State for War issuing the orders. Stephen
Long on his reconnaissance to the Missouri was thus charged to 'enter in
your journal everything interesting in relation to soil, face of country,
water courses and productions, whether animal, vegetable or mineral'.
The records of these early military explorers are among the treasured
documents of the Library of Congress in Washington. They include the
maps and notebooks from the expedition of Meriwether Lewis and
William Clark to the Trans-Mississippi lands (1803-6), the reports of
John Frémont from the Oregon trail and California surveys (1843-4)
and the War Department explorations for a railroad to the Pacific

(1854–7). The great Western Surveys of the inter-montane lands (1867–79) yielded a harvest of geological materials, panoramas and profiles. To this period also belongs the first use of the camera in reconnaissance; though Captain Simpson's engineers were not very enthusiastic about the 'photographic apparatus' on their expedition into Utah territory. There were pathfinders in Canada as well. Those who first recorded the character of the land have often given their names to its features. In the perspectives of history men such as Alexander Mackenzie and John Palliser seem to have belonged to a race of giants.

The spread of the official survey and the age of exploration ran side by side, but during the final phases of North American settlement, survey lagged behind occupation of the land. The rectilinear grid which forms the basis of survey for the public lands of North America had its origins in a land law conceived by Thomas Jefferson and Hugh Williamson in 1784. There was some debate concerning the size of the townships which were to be its basic units, but it was eventually decided in 1796 that they should be of six miles square and that the grid should conform to true north. To Thomas Hutchins, nominated geographer in charge of surveying in 1785, may be ascribed the methods of record-keeping adopted in the field. The grid was born on the banks of the Ohio and the present-day map reflects an essential contrast between the irregular pattern of settlement features east of the Appalachians and the regular pattern which predominates to the west.

The tradition established by Hutchins was maintained by his successors and, as a result, a notable record of the original scene has been retained in the surveyors' field notebooks. State archives or county-record offices usually keep the field notebooks for their own areas, so that the geographer who would look into the original occupation of lands beyond the range has a unique body of material available. On the map, and usually on the land as well, east-west township lines with their intervening concession lines are intersected by north-south range lines. The six-mile square was subdivided into thirty-six sections of one square mile each. The numbered section has been the basis of the land-holding system. Urban settlements, as well as rural areas, generally display a grid pattern, this formal civic order already being established in the eighteenth century (Figure 1).

Although the topographical mapping of Canada has taken different forms from that of the U.S.A., there is similarity in its township surveys. The stages of land acquisition have altered little since township surveys were established in the 1790's. Measurement was by surveyor, with

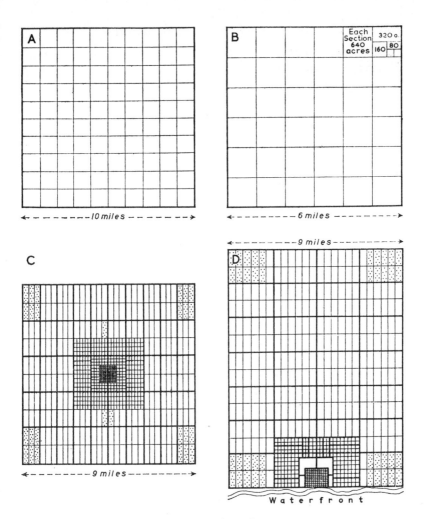

Figure 1 *Township and Range*

The Jefferson proposals for a ten-mile square township (A) were relinquished in favour of a six-mile square township (B), with numbered lines of division and primary units of a section. The same method of land division was transferred to Canada and two formal plans are reproduced from A. Fraser, *16th Report of Department of Archives for Province of Ontario*, Toronto, 1921. (C) is a plan of a town and township proposed for 'an inland situation agreeable to the tenth article of the rules and regulations for the conduct of the Land Department of the 17th of February 1789 by order of His Excellency Lord Dorchester'. (D) is a similar plan 'of nine miles front by twelve miles in depth proposed to be situated on a river or lake' according to the same authority 'chain bearers and axemen . . . running the lines' through the countryside and keeping a survey diary, field notes and transcript map. Farm lots of 200 acres (80 rods frontage and 400 rods deep) were allocated to colonists along concession roads which were 40 to 60 feet wide. On occupying a farm lot a settler acquired a record of location, while a land patent was given when he had complied with settlement conditions. A specific example from the Ontario waterfront, the Certificate of Settlement for Concession IV, South Dundas Street, Nelson Township, Broken Lots 4 and 5, ran as follows:

> Before me Alexander Woods Esq., one of H.M. Justices of the Peace for the Home District appeared Phillip Triller, William Triller both of Nelson yeomen and made oath that there are 5 acres cleared and well-fenced on the broken lots 4 and 5 . . . and that a house of 18 feet by 20 feet is erected on the said lots and that the road is cleared in front of the said lots. Sworn before me at York this first day of August, 1809.

A 'broken lot' is a land unit which is not squared along one or more sides, in this case on the south by the coast of Lake Ontario. The Certificate of Settlement made reference to a Plat Doomsday, or Doomsday book of plots. The Provincial Archive in Toronto houses the so-called *Doomsday Book of Ontario*, which is arranged by townships and records all land grants. The historic procedure of allocating virgin land persists today, whether it be for summer cottages and fishing cabins in up-country Ontario or for entire farmsteads in 'colonial' Canada. Town lots have been drawn up in a parallel way. Figure 1 illustrates sample outlays for two townships, one facing a watercourse.

Township surveys have been superimposed upon the face of the land virtually regardless of topographical features. Maps for such contrasting tracts of country as Kirkland Lake, Ontario (Figure 57), and Vancouver B.C. (Figure 62) illustrate the discomfiture with which the frame of township and range sits upon topography. In Coquitlam it results in a kind of meridional madness in which the third dimension is ignored: in the amphibious Kirkland Lake area Roman-straight divisions are sought in a landscape which cries out for the corkscrews of the English road.

Responsibility for the topographical mapping of Canada, first undertaken by the British Ordnance Survey (which was still printing its Canadian series in the Tower of London less than a century ago), rests jointly with the Army and the Department of Mines and Surveys in

Ottawa. Topographical maps on a scale 1:63,360 or larger cover only a fraction of Canada; but the entire country has been photographed from the air, so that economically or strategically important areas can be photogrammetrically mapped as and when the need arises.

In the U.S.A. the Geological Survey of America has been responsible for issuing large-scale maps. The Survey was formally established in 1879 and as a result of its energies the U.S.A. can claim to have the most extensive coverage, on a scale 1-25,000 or greater, of any country in the world. But the size of the U.S.A. implies that for only half of it are there detailed topographical surveys. Today, many varied agencies—e.g. the Soil Survey, Water Resources Board, Conservation Survey—are at work recording the land. In addition to its maps the Geological Survey has produced a great library of monographs and memoirs. These constitute an invaluable body of raw material to which the physical geographer may turn for his understanding of the land of America.

GEOMETRY IN NORTH AMERICAN GEOGRAPHY

The geography of North America is in large measure a function of its shape and position (Figure 2). The continent is roughly triangular in outline with corners at Bering Strait, latitude 66°N, longitude 170°W; Newfoundland 48°N, 53°W; and the southern border of Mexico, 15°N, 92°W. The broad base is in the north and the apex points to the equator. It is interesting to speculate how the geography of the continent would differ were the triangle inverted. The coasts are not straight; gulfs and bays bite into the sides of the triangle and peninsulas protrude beyond them. About half the land surface at the apex is replaced by the sea surface of the Gulf of Mexico, although the Florida peninsula partially compensates by protruding south-eastwards. The base is broken by Hudson Bay, but this is counterbalanced by the arctic 'peninsula' made up of the islands between the north coast of Canada and Ellesmere Island. The Gulf of St. Lawrence and the Great Lakes are smaller but extremely important penetrations of the land by water. The Aleutian peninsula and islands point a long finger from the Alaskan corner towards Asia. Between Alaska and the Columbia river estuary the coast swings well to the east of the straight side of the triangle. Compensation is found in the complementary bulge south of the Columbia at the apex of which Cape Mendocino, 41°N, marks the western limit of the forty-eight states. It is roughly along this line of latitude that the United States attains its maximum east-to-west extent. The Gulf of California and its mirror image, the peninsula of Lower California, complete the picture.

The very size of the continent has been an important factor in its human exploitation. The longitudinal extent of the triangular base stretches through 116° and the latitudinal extent is 68°. The continent centres upon two axes—one north-south between 90 and 100°W and the other east-west—between 40 and 49°N. These particular lines of longitude and latitude have considerable geographical significance: 90°W is the

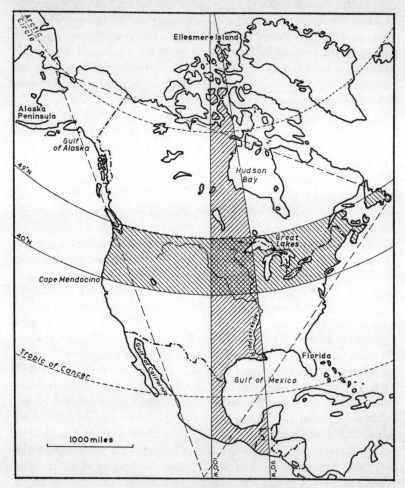

Figure 2 *North American Geometry*

central axis of the Mississippi Valley; 100°W represents an important if uncertain boundary between the humid eastern U.S.A. and the dry west; 49°N is followed by the U.S.-Canadian border; whilst 40°N, as indicated, is approximately the line of maximum east-west continental extent in the U.S.A. In the west 40°N is also the approximate northern limit of the Mediterranean type of climate.

The 7,602,617 square miles of land area are shared very unequally by three countries: Mexico 760,375 square miles, the U.S.A. 2,997,128 square miles and Canada 3,845,114 square miles. This book is concerned only with the last two. Within their borders are great diversities of landscape which have developed and are still developing as a result of interaction between people of many origins living on a land of varied character. It is this aspect which now concerns us: the diversity of physical background can best be understood by an analysis of its rocks and soils, relief and drainage, climate and vegetation.

THE ROCKS AND THEIR ARRANGEMENT

The structural core of the continent is the expanse of Pre-Cambrian or Archean rocks of the Laurentian Shield, so called because of its overall shape (Figures 3 and 4). This is not at the centre of the continent as defined by the intersection of the principal axes on Figure 2 but in the north-east quadrant. Except for the Superior Upland and the Adirondack Mountains it lies entirely within Canada, hence the frequent appellation Canadian Shield. Modern geological exploration, building on the solid foundation of knowledge acquired in the years from 1843 to 1939, has greatly extended what is known of the shield. The search for gold, silver, copper, nickel, iron and uranium has been the stimulus to geological exploration of this difficult terrain.

The surface rocks are, for the most part, highly folded and metamorphosed; though in certain areas, notably on the southern shores of Hudson Bay and around Lake St. John in the east, sedimentary rocks of Ordovician age rest unconformably upon the Pre-Cambrian rocks.

This structural heartland extends to the south and west below the present surface in the Central Stable Region, where it is covered by a wedge of sedimentary rocks, including sandstones, shales and limestones such as the Indiana freestone which is widely used as a building stone for public buildings in the U.S.A. With this solid foundation, the region, as its name implies, has been able to withstand subsequent earth movements

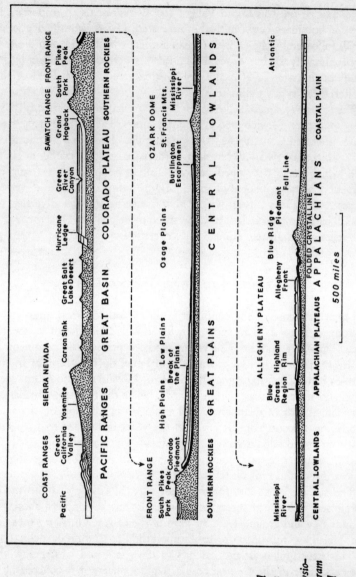

Figure 3
A Geological
Cross-section
(after A. K.
Lobeck, Physio-
graphic Diagram
of the United
States, New
York, 1948)

and the overlying rocks have been only gently folded into arches and basins. East of the Mississippi the rocks are of Palaeozoic age and include the principal coal-bearing strata. The latter are preserved in the basins which are synonymous with the chief coalfields of the U.S.A. The chief upfold is the Cincinatti Arch, which comprises the Nashville and Lexington domes. West of the Mississippi the Palaeozoic veneer is in turn covered by gently dipping Mesozoic and Tertiary rocks derived from the

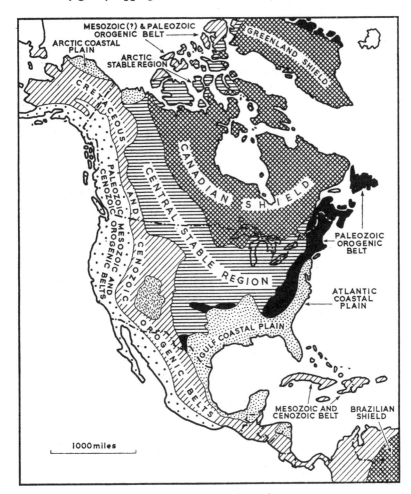

Figure 4 *The Geological Foundations*
(after A. J. Eardley, *Structural Geology of North America*, New York, 1951)

erosion of the Cordillera to the west. In these rocks are found the principal oil pools of the continent.

Welded on to the core of the continent along its eastern and southern margins is the Palaeozoic orogenic belt. The present-day Appalachian ranges, extending south from Newfoundland to Alabama, are but the worn-down and rejuvenated remains of the mountains formed by these early orogeneses. Here, unlike the Central Stable Region, the Palaeozoic sediments have been intensely folded, thrust and faulted. It is usual to distinguish between the inner (so-called Younger) folded Appalachians which extend from Alabama to New York State, the rocks of which are sedimentary in origin and contain no volcanic elements, and the outer (so-called Older) crystalline Appalachians, made up of a considerable variety of volcanic rocks including granite batholiths and trap rocks, that is, lavas. The southerly extension of this belt is broken into a series of isolated regions where the coastal plain sediments overlap the Central Stable Region.

The next period of mountain building produced the orogenic belt along the Pacific margin and the process has not yet ceased. The belt is volcanic, as witness the extinct volcano of Mount Rainier and the still active (1922) Lassen Peak. Between it and the Stable Region there existed for a considerable period a trough of deposition or geosyncline, which was folded in the Laramide orogeny to form the third or Rocky orogenic belt. This formation took place at the end of Cretaceous times and the erosion of these folds yielded the material which was carried eastwards and deposited in the western part of the Central Stable Region. In the U.S.A. the folds divide into two branches enclosing a median mass where the rocks are still relatively horizontally disposed, forming the Colorado plateau province.

In the south and east, from Cretaceous times onwards, subsidence has occurred periodically and sedimentary rocks have been deposited to form the Coastal Plain. The sands, marls and limestones have a seaward dip but have been gently uplifted in the Florida peninsula.

THE HEIGHTS OF LAND AND THE WAYS OF THE WATERS

While these structural elements form the basis of a division into land-form regions it is not only rocks, folds and faults which determine the shape of the ground. Erosion and deposition, particularly by rivers and ice in the recent geological past, have been responsible for the third-order relief features.

The surface water from the North American continent finds its way, sometimes by devious routes, into three oceans—the Atlantic, Arctic and Pacific in order of area drained (Figure 5). The watershed between the Pacific and Atlantic Oceans, the Continental Divide, follows the axis of the Rocky Mountains, so that between two thirds and three quarters of the U.S.A. is drained east and south to the Atlantic. The divide is sometimes a formidable obstacle, as where U.S. 40 has to climb to over 11,000 feet to cross it west of Denver (Col.), at other times low or

Figure 5 *Water on the Land*

C

punctured by passes, e.g. south of Albuquerque (New Mexico), which is the route followed by U.S. 66 and the Sante Fé railroad. In Wyoming the Union Pacific railroad follows the broad sag of the Wyoming Basin between the Colorado Rockies and northern Rockies in the Yellowstone and Grand Teton National Parks. Narrower defiles across the Continental Divide are the Crow's Nest (3,500 feet), Kicking Horse (5,329 feet) and Yellow Head (3,729 feet) passes. The two latter are followed by the transcontinental routes of the C.P.R. and C.N.R. railroads respectively. The Kicking Horse pass is also used by the Trans-Canada highway. From the 49th parallel to a few miles north of Mount Robson the divide forms the provincial boundary between Alberta and British Columbia. North of this, British Columbia includes that often forgotten part of the Great Plains region west and south of 120°W and 60°N, and the divide swings west around the headwaters of the Peace and Mackenzie river systems which head back beyond the Rockies into the Stikine Mountains, west of the Rocky Mountain trench. The Alaska Highway crosses the divide in a pass between the heads of the Liard and Teslin, tributaries of the Mackenzie and Yukon respectively. The divide then makes an eastward bend, following the Mackenzie Mountains around the headwaters of the Yukon River before heading westwards again, close to the Arctic shore, along the axis of the Brooks Range, the termination of the Rocky Mountain system.

The Pacific drainage is extremely varied in pattern and régime as befits its wide latitudinal spread. The Yukon drains the plateau area between the Alaskan coast ranges and the Rockies into the Bering Sea. It is a semi-arid area with an annual precipitation of less than 20 inches and sub-zero winter temperatures. To the south much shorter rivers, such as the Juneau and Skeena, drain directly to the Pacific in regions where mean annual precipitation is 80 inches and over with little seasonal variation. All are nourished in part by winter snows at their sources; some, by permanent ice caps. The Fraser and Columbia systems drain from the Rockies across the grain of the country through the Cascade and Coast mountains, on the one hand fed by orographic rains which decrease in amount eastwards with each successive longitudinal mountain range and on the other sapped of their strength by evaporation from river and lake surfaces in the rain shadow troughs between the ranges. South of the Snake, the Great Basin is a region of inland drainage where mean annual rainfall is less than 10 inches p.a. and temperatures exceed 100°F in the shade in Death Valley. Rivers such as the Humboldt are fed by short streams from the flanks of peripheral mountains, for example the Sierra

Nevada in the west and the Wasatch range in the east. Others, fed from the fault block mountains within the basin, either lose themselves in the playa floors of the desert or form saline inland waterbodies such as the Great Salt Lake. Only the orographic rain-and-snow-fed allogenic Colorado survives the journey across the desert lands from the Colorado Rockies, the San Juan and Unita mountains to the Gulf of California. The régimes of the Sacramento and San Joaquin rivers, which drain the north and south lobes of the Californian central valley, well portray the climatic differences between their basins. The former is on the northern margins of the Mediterranean type of climate, the latter has a hotter and drier version of that climate, much influenced by rain 'shadow' effects. Both are well fed by streams rising in the Sierra Nevada, but the Sacramento also has tributaries from the coast ranges. The Kern, which might have been the source of the San Joaquin, fails to reach it in the desert outlier around Bakersfield, at the southern extremity of the Great Valley.

The roughly west-to-east divide between the Atlantic-bound streams and those draining to the Arctic Ocean, including Hudson Bay, is much less well marked in the landscape than the Continental Divide. No high physical barrier protects the continental interior from the climatic influences of the Arctic—in marked contrast to the situation in Europe where the Hercynian and Alpine mountain systems set a southward limit to continental climatic characteristics. Between the Saskatchewan and the Missouri the divide crosses the higher steps of the Prairies and the 49° parallel into the U.S.A., where it bulges south to include the Red River Valley, a southern extension of the basin of the former pro-glacial Lake Agassiz. It then passes north around the source of the Mississippi, Lakes Superior and Huron (where it runs along the rim of the Canadian Shield), and north-eastwards along the axis of Quebec province. From a short distance south of Schefferville, as far as Cape Chidley, it is followed by the boundary between the province of Quebec and Newfoundland (Labrador).

Of much greater significance are two divides south of this. The first separates the rivers bound directly for the Atlantic from those flowing to the Gulfs of Mexico and St. Lawrence. The second separates the Great Lakes and the Mississippi-Ohio system. The former starts in the south on the axis of the Florida peninsula amidst swamps and sink-hole lakes. In the coastal plain of Georgia it crosses Okefenokee Swamp, the source of Stephen Foster's Suwannee River and the cartoon home of satirically minded Pogo and his animal friends. In North Carolina the divide follows the Blue Ridge along the eastern margin of the Appalachians to heights of 6,684 feet in Mount Mitchell. In the Virginias it moves obliquely north-west across the grain of the ridge-and-valley country, between the east-bound Roanoke and the headwaters of the Tennessee, one of the routes of pioneer settlers. Around the sources of the Potomac and Susquehanna in Maryland and Pennsylvania the divide runs first on the crest of the Allegheny Scarp and then, farther west, on the Allegheny plateau. The most impressive breaches in the divide between the Delaware and the Gaspé peninsula are the Mohawk Valley, followed successively by Indian trail, Erie Canal, New York Central railroad and super highway; the Champlain Valley route to Montreal between the Adirondacks and Green mountains; and two less spectacular Canadian cols used by the trans-continental routes of the C.P.R. and C.N.R.

It would require very little tilting of the Great Lakes to make their waters spill southwards across the low divide of glacial drift into the tributaries of the Mississippi, as they have on more than one occasion in their brief history. Old overflow channels at the head of the rivers St. Croix (Lake Superior), Illinois (Lake Michigan) and Wabash (Lake Erie) form short portages between these important waterways to give easy entrance to and exit from the human heart of the continent.

A SURVEY OF THE MORPHOLOGICAL REGIONS

One of the most striking features of the U.S.A. and Canada is the clarity with which their morphological regions can be distinguished (Figure 6). Since 1918 there has existed a definitive nomenclature for these regions of the U.S.A., based upon the work of N. M. Fenneman. In order to give some impression of these contrasting regions it is proposed to describe two transverse sections of the continent. The first is of the unglaciated lands and therefore concerned with the U.S.A. The second, largely but not exclusively Canadian, is through the glaciated lands. The two will be linked by a consideration of the effects of the Pleistocene glaciations.

I. A CROSS SECTION OF THE UNGLACIATED LANDS

The Atlantic Coastal Plain (Region 1a on Figure 6)
South of New York the Atlantic coast is bordered by a coastal plain which broadens progressively southwards. The present coastline is an arbitrary line dividing two halves of a great wedge of relatively young and un-

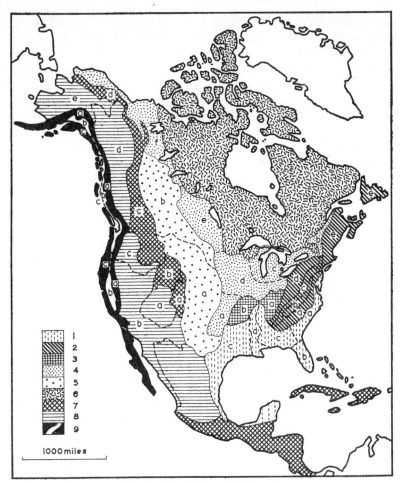

Figure 6 *The Shape of the Land*

1 Coastal Plains. 2 Appalachia. 3 Interior Uplands. 4 Interior Lowlands. 5 Great Plains. 6 Laurentian Shield. 7 Rocky Mountains. 8 Inter-montane Plateaux. 9 Pacific Mountains (after A. K. Lobeck, *Physiographic Provinces of North America*, New York 1948). See pp. 36–68

consolidated rocks. These rest upon an older basement of crystalline rocks exactly as in the Central Stable Region, although the rocks involved are of different ages. That part of the wedge where the upper surface is above sea level forms the Coastal Plain, that part submerged beneath the waters of the Atlantic is the Continental Shelf. The rocks are mainly sands, gravels, clays and marls, with local developments of limestone. They dip gently eastwards, but because of their unconsolidated character do not form a pronounced scarp and dip-slope topography. The diversification in the relief is provided by low hills near the inner margin and broad, flat, low plateaux surfaces separated by bluffs, 'terraces' and 'scarps', as they are misleadingly called, e.g. the Surrey Scarp. The origin of these plateaux and bluffs is not agreed upon, but they are most probably the work of the sea. They may be beaches and low cliff lines, cut in the intervals between successive glaciations at times when sea level stood higher than at present. Spread over these old sea floors are superficial deposits of sand and gravel laid down both in the interglacial seas and by rivers which extended themselves across the emergent coastal plain. Some of the lowest plateaux are still without an integrated drainage pattern and are therefore swamps. The rivers have cut into this sandy, still extensively wooded, terrain so that they are bordered by valleyside bluffs, but their gradients are low and they have, by meandering, developed broad marshy bottom lands. The last major change of sea-level uplift, following the final or Wisconsin period of glaciation, drowned the lower ends of these valleys back as far as the Fall Line, particularly in the northern or embayed section of the coastal plain. Chesapeake Bay, with the estuaries of the Susquehanna, Potomac and James rivers, is a yachtsman's paradise. Albemarle Sound and Delaware Bay are the drowned lower ends of the Roanoke and Delaware, on whose floors can be traced the former sub-aerial channels. Bordering the coast are sand-bars and beaches, the site of many a summer cottage and the setting of not a few seaside resorts.

South of Cape Hatteras barrier beaches and lagoons are more frequent, drowned valleys less frequent. Inland there are a considerable number of shallow oval lakes, the Carolina bays, variously attributed to meteoric, submarine, aeolian and permafrost origins.

Strictly speaking, the coastal plain ends south of Staten Island, but, together with Long Island and Martha's Vineyard, Staten Island forms a disjointed topographic continuation. The deposits of which they consist are remnants of two great terminal moraines, the Harbor Hill and Ronkonkoma, and they mark the southernmost extension of the ice on the eastern seaboard.

The Florida Peninsula (Region 1b)

Florida is an uplifted extension of the coastal plain in which limestone is more important than elsewhere. The centre of the peninsula, rising in places to over 200 feet, is riddled with sink-holes which sometimes occupy more than half the surface. They are almost all partly filled with water. The watertable follows closely the treads in the staircase of old sea floors between which former cliff lines are traceable for mile after mile across the forested landscape. Phosphate is an economically important component in the rock sequence near Mulberry. In southern Florida swamps only a few feet above sea level occupy several thousands of square miles; the best known are the Everglades. The Atlantic coast is fringed by swamps, lagoons, barrier beaches and spits. They are havens for those who can afford to fly from the rigours of the north-eastern winters and for those who wish to make rockets and fly still farther into space; Palm Beach and Cape Canaveral have that much in common. The curved line of islands called the Keys are based upon old coral reefs.

The Piedmont (Region 2a)

The well-marked junction between the sands and clays of the coastal plain and the crystalline basement complex of the Appalachians is expressed topographically as the 'Fall Line'. The eastern edge of the crystalline rocks is being stripped clean of its cover of sedimentary rocks to form a bluff. This is especially clear where a river follows a subsequent course along the junction, so that one side of its valley is the Fall Line bluff and the other the innermost scarp of the coastal plain. Between Trenton and Philadelphia the lower Susquehanna Valley follows such a course. Irregularities in the long profiles of the rivers where they cross the Fall Line are now frequently some distance upstream of the junction between the two rock types as a result of the retreat of the knick point, e.g. the Great Falls on the Potomac above Washington.

The Fall Line is the easternmost limit of the crystalline Appalachians, which comprise two parts: the Piedmont and Blue Ridge. These differ little in rock character but markedly in relief. The Piedmont is a surface of rolling relief, a peneplain worn across Pre-Cambrian metamorphosed rocks rising from heights of 200 to 500 feet at its outer margin, to 1,000 to 1,200 feet inland, giving an average gradient of 10 feet per mile. There are occasional monadnocks rising above it, such as the particularly striking Stone Mountain in Georgia. In the south, monadnocks increase in number towards the inner edge and form a transition belt of hills to the Smoky Mountains and the Blue Ridge. The whole belt broadens from a mere 30 miles in New Jersey and Pennsylvania to 150 miles in North Carolina. Within the high Piedmont terrain there are lowland tracts developed upon less resistant Triassic marls and limestones which have been down-faulted into the Pre-Cambrian rocks. They weather to a better quality soil than the metamorphic rocks. Such lowlands are found from Georgia to New Jersey. In New Jersey the Triassic rocks include lava flows (trap rocks) bedded with the sedimentary rocks. Their uptilted edges form striking trap ridges, such as the Watchung ranges, whose scarp edges rise a few hundred feet above the surrounding lowlands. The most famous is the dark rock wall of the Palisades bordering the Hudson River on the Jersey shore, from the crest of which are obtained the finest views of the Manhattan skyline.

The Blue Ridge (Region 2b)

In the south the Great Smoky Mountains form a mountain knot some 70 miles wide, which rises to elevations of over 6,000 feet. From them drain the headwaters of the Tennessee, which rise at points overlooking the Piedmont and then flow westwards across the mountains. Along their courses considerable lowland areas are developed, such as the Ashville Basin, the floor of which is a peneplain at approximately 2,000 feet elevation. The forested ridges continue northwards as the Blue Ridge, broken by the deep and narrow water gaps of the eastward-flowing Roanoke, James and Potomac rivers, as well as by numerous waterless gaps and cols at a great variety of elevations. Some of these may be wind gaps marking the sites of transverse rivers, later captured by the development of subsequent streams in the weaker rocks west of the Blue Ridge. Thus, Manasas Gap in Virginia lost its stream through piracy to the Shenandoah. In Pennsylvania the Blue Ridge fades away almost completely and no barrier separates the Piedmont and the valleys beyond. At this point the Susquehanna, fortuitously or otherwise, crosses the crystalline Appalachians and has found no obstacle in its path. When the Blue Ridge picks up again at Reading (Pa.) it is as the Reading prong which may be looked upon as a finger pointing south-westwards from the broad palm of the New England version of the crystalline Appalachians. At the knuckle the Hudson has cut a water gap through the Hudson heights, now the site of West Point.

The Ridge and Valley Province (Region 2c)

West of the Blue Ridge the rocks are mainly sandstones, limestones and shales of Palaeozoic age, folded in late Palaeozoic times into a series of concertina-like structures in which folds are traceable for many miles before they pitch and disappear. The forces of erosion have cut down into this folded pile of alternately resistant and less resistant rocks to produce a landscape in which long sandstone ridges and shale and limestone vales alternate in rapid sequence. The complexity is increased by the replacement of folds by faults and thrusts, but is sometimes simplified by a local absence of folding. Depending upon depth of erosion, the ridges may be coincident with anticlines, the vales with synclines. The relief may also be inverted, and hogsback ridges circumscribe anticlinal vales overlooked by synclinal mountains with outward-facing scarps.

The easternmost vale, which is next to the Blue Ridge, is also the broadest. This is the Great Valley, developed upon a repeated outcrop of limestones. It is traceable from Alabama to New Jersey, and is made up of the valleys of a number of subsequent rivers, among which are the Coosa, Susquehanna and Kittatiny. As a routeway and favoured settlement site, it has played an important role in the historical geography of the U.S.A. Its floor is dissected into a number of broad terraces which have been recognized as important stages in the geomorphological development of the Appalachians. The best known is the Harrisburg peneplain, 500–600 feet, named after that part of the Great Valley at Harrisburg (Pa.), in the Susquehanna Valley.

Beyond the Great Valley the ridges follow one another in rapid succession. There are strikingly named rocks, evocative of Indian times, such as the Oriskany, Pocono and Tuscarora sandstones separated by vales on the Chemung and Mauch Chunk shale. To drive across them on the minor roads at right angles to the grain of the country is to experience a roller-coaster ride in slow motion. In the south, where thrusting is more important than folding, the ridges are less numerous and lower. Major streams cut across the even-topped ridges in spectacular river gaps; a feat performed by the Susquehanna five times in less than ten miles upstream of Harrisburg. On the other hand, there are many subsequent streams, such as the Tennessee, which follow the grain of the country. Together they form a trellis drainage pattern, which is a natural corollary of the type of relief. Many ridges are flat-topped; but others, which are not, also accord with them in height. Such features provide evidence of the much-dissected, uplifted Schooley peneplain. The transverse nature of the major rivers is variously explained as having been superimposed across the structures or resulting from headward erosion through weak points in the ridges.

In Pennsylvania, around Wilkes Barre, coal-bearing rocks outcrop in the cores of a series of tight synclines. As a result of the compression, the coal has been transformed into anthracite.

Northwards, the folding becomes less intense. From a maximum width of 80 miles in Pennsylvania the fold belt narrows rapidly in New

York State, where the Great Valley is continued by the Hudson Valley between Newburgh and Albany, and beyond along the Champlain route. The Catskill Mountains, a highly dissected portion of the western plateau, almost replace the remainder of the Ridge and Valley belt.

Southwards, in Alabama, the vales between the ridges lead out, unhindered, to the coastal plain. The drainage of the southern Appalachians is directly to the Gulf of Mexico via the Coosa River, a tributary of the Alabama. The well-developed Coosa peneplain is probably the equivalent of the Harrisburg surface farther north. Its broad sweep is diversified by linear monadnock ridges.

The Appalachian Plateau (Region 2d)

The folds of the inner Appalachians die out to the west, and the Appalachian plateau is developed on a broad synclinal trough trending northeast to south-west from the Mohawk to beyond the Tennessee Gap. It narrows considerably in the south, where it is crowded in by the marked dome structures of the Cincinnati Arch immediately to the west.

The whole of the Palaeozoic rock-pile is represented in this province, including the coal-bearing Pennsylvanian (Carboniferous) sediments, which are sufficiently high in the pile to have been preserved along the axis of the syncline. Minor anticlinal folds, the outer ripples of the Appalachian orogeny, diversify the synclinal structure and determine the distribution of oil deposits in the Pennsylvanian rocks.

The even surface of the plateau is a peneplain (perhaps a series of them) bevelling the synclinal structure. Above it rise ridges, such as Chestnut and Laurel ridges developed on the flanks of minor anticlines. In southeastern Kentucky and Tennessee there are a number of these folds as well as some thrusts, so that the distinction between ridge-and-valley and plateau is largely one of the relative importance of ridge or valley. The ridges are broader and higher and the valleys narrower in the plateau than in the Ridge and Valley Province. The average height of the plateau is 1,200–2,000 feet. It is delimited fairly precisely around its whole perimeter. In the east the bold scarp edge overlooks the inner vales of the folded belt. The formidable rise is known in Pennsylvania as the Allegheny Front or Allegheny Mountains. Two railroads, the Baltimore & Ohio and the Pennsylvania railroad, negotiate this scarp in a series of spectacular horseshoe bends, both major feats of engineering in their day and still an experience to ride. The term Cumberland Front is applied to the same scarp in the south. It is broken by a narrow high-level valley, the Cumberland Gap, a piece of geomorphology and historical geography perpetuated in popular song.

The plateau is divisible into a number of distinct units. In the north the Catskills, developed on more resistant Devonian deltaic deposits, have been maturely dissected and glaciated. They rise above the Schooley peneplain by some 2,000 feet and tower above the adjacent Hudson Valley by some 3,000 feet. The glaciated New York section terminates in the Helderberg escarpment which forms the south side of the Mohawk Gap. Farther west, the north-facing fringe scarp was deeply dissected by ice which moved southwards against the regional slope up the scarp, cutting the prodigiously deep troughs now occupied by the Finger Lakes. This northern part of the plateau drains to the Susquehanna. At the valley heads there are a number of cols leading across the water parting to the Great Lakes system, the sites of melt-water overflow in the early stages of deglaciation and the formation of the Great Lakes.

In Pennsylvania the Allegheny and Monongahela rivers deeply dissect the plateau surface and the horizontal strata beneath. The numerous bituminous coal seams (and, formerly, iron-ore beds) are thus exposed to easy adit mining. Terraces and occasional abandoned channels lent themselves as sites for settlement, communications and industry. At the confluence of the rivers which form the Ohio lies Pittsburg, the focus of this geologically rich but geomorphologically restrictive environment. The coal measures continue into West Virginia, where the Kanawha, which rises on the other side of the Ridge and Valley Province in the Blue Ridge, is incised some 1,000 feet below the plateau and is followed by the Chesapeake and Ohio railroad. The Tennessee at Chattanooga turns from its subsequent course, crosses the Walden Ridge into the Cumberland plateau, flows again to the south-west and finally crosses the remainder of the plateau, north of the Warrior Basin, in a river gorge south of Scotsboro.

The Interior Uplands (Region 3)

West of the Appalachians is a vast lowland stretching from the Gulf of Mexico to the Arctic Ocean. In a traverse from the mouth of the Mississippi to Hudson Bay there is no need to exceed a height of 1,000 feet. In the Mississippi Valley stand two upland masses, one each side of the embayment of the coastal plain which extends northwards along the Mississippi to its junction with the Ohio at Cairo (Ill.). The eastern upland is referred to as the Interior Low Plateaux (3a). The gentle syncline of the Appalachian plateau gives place westwards to the arches of the Lexington and Nashville domes. Erosion by the Kentucky and Cumberland rivers has unroofed these structures and produced two topographic basins: the Kentucky Blue Grass country in the north and the Nashville Basin in the south. These are encircled by concentric scarps which rise to an upland plateau or highland rim itself overlooked in the east by the Appalachian plateau.

The floor of the Nashville Basin is some 500 feet below the highland-rim plateau, which slopes westwards from 1,200 feet in the east to 700 feet along the Tennessee River in the west. The lower Tennessee Valley roughly outlines the southern and western limits of the Nashville Dome. The eastern highland rim is called the Barrens. North-west of the Nash-ville Basin is the Pennyroyal plateau. The principal scarp formers are thick sandstones, the edges of which are often dissected into a distinctive terrain of conical hills known as the 'knobs'. The plains are underlain by lime-stones. The phosphatic limestones of the Inner Blue Grass are the physical basis of the Kentucky Colonel, his whisky and the Kentucky Derby. The Outer Blue Grass is more rugged than the rolling true Blue country. The karst is best developed on the Pennyroyal plateau, where square mile after square mile is pitted by hundreds and sometimes thousands of sink-holes. Beneath the plateau surface are cave systems of impressive magnitude. The best of them, such as Mammoth Cave, are found where the limestones are buried beneath the sandstones of the Dripping Springs escarpment. Karst valleys are numerous where the overlying sandstone has been dissected to reveal the limestones beneath. There are also many examples of cave collapse and, where this has led to the collapse of the overlying sandstones, some of the largest holes are formed. Disappearing streams, underground drainage, blind valleys, dry valleys (as many on the sandstones as on the limestones) proliferate.

In southern Indiana the alternating shales, sandstones and limestones dip westwards beneath the coal deposits of the East Central Coalfield. Dissection of an upland plain has produced a series of east-facing sandstone scarps alternating with lowlands on shales or limestones. The Ohio after skirting the northern flank of the Cincinnati Arch cuts through these scarps and separates the Western Kentucky coalfield from its Indiana and Illinois counterparts.

The Ozark uplift (3b) west of the Mississippi and south of the Mis-souri is a mirror image of the Nashville Dome. But here erosion has ex-posed part of the granitic floor in the St. François Mountains, the relative relief of which is only some 800 feet above the surrounding land. The Palaeozoic sedimentary rocks dip gently away from the apex of the uplift and their upper surfaces have been peneplained to form extensive plateaux. The Salem Upland is a plateau on which karstic landforms are developed in the dominant cherty and dolomitic limestones, between narrow sandstone outcrops. The gentle dip is westwards to the Burlington escarpment and southwards to the Eureka Springs scarp, which marks the edge of the Springfield plateau. This is the site of the Tri-State lead and zinc deposits, the principal source in North America. The ores are contained within rocks of Mississippian (Lower Carboniferous) age. In the extreme south is the true Ozark country of the east-west oriented Boston Mountains, in reality a highly dissected plateau on Potsville sandstones of Upper Carboniferous (Pennsylvanian) age with summit heights of over 2,000 feet. In the north a series of scarps step down to the Springfield plateau and southwards there is an abrupt slope down to the Arkansas Valley in conformity with a monocline which takes the sandstones down under the valley before they rise again on the Ouachita Appalachian type folds to the south.

The principal rivers of the Ozarks are the Osage and Gasconde, which drain the Springfield-Salem upland north to the Missouri; the White which drains the Boston Mountains eastwards to the Mississippi; and its tributary the Black, draining the St. François Mountains. These rivers, through their powers of down-cutting, are creating a mountain area out of a staircase of plateaux. The adjacent Arkansas Valley lowland is cut in a broad belt of coal-bearing shales which occupy the core of a synclinorium south of the Ozark uplift. Sandstone beds outcropping amongst the shales form low east-west ridges. Traversing the length of the syncline is the broad alluvial plain of the Arkansas River which encounters the rocks of the Mississippi embayment at Little Rock. South of the synclinorium is the complementary upfold of the Ouachita Mountains. They consist of homoclinal ridges and vales arranged in a zig-zag pattern as befits the work of streams upon a tightly packed series of parallel, pitching folds. The maximum altitudes reached on the principal ridge-former, which is a much-quarried chert, are about 1,500 feet. The trellis drainage pattern also reflects the same structural influence.

The Gulf Plain Coast and Mississippi Embayment (Regions 1c, 1d)
The gently dipping rocks of the coastal plain wrap around the open ends of the Appalachian system in the East Gulf Coastal Plain region. It is a true belted coastal plain comprising a parallel series of cuestas and intervening lowlands. The most prominent ridges are the Red and Pine hills. The innermost lowland is the Black Belt of Alabama developed on the Selma

chalk. There is a short extension of the limestone country into this region from the Florida peninsula. The concentric outcrops are crossed by extended consequent streams such as the Chattahoochee, Alabama and Tombigbee which drain from the old land of the Appalachians across the upraised sea floor. The Chattahoochee reaches the sea as the Apalachicola River through a considerable delta, the others enter Mobile Bay in lagoons.

The Mississippi embayment is a northward extension of the coastal plain whose axis is the broad lowland of the meandering Mississippi River below the fork of the Ohio. The alluvial plain, from 50 to 100 miles wide, is by no means devoid of geomorphological detail. A complex pattern of low ridges and shallow, curved depressions marks the natural levées and river beds of abandoned meander loops. These flood plain scrolls and bayous have soils which differ markedly in texture and drainage characteristics. The scrolls have coarse and well-drained soils; the bayous, clayey and wetter soils (Figure 38).

There is also variety on a larger scale, for the course of the river has changed repeatedly. At one time it flowed west of the Rowley ridge, a valley now followed only by waters of Ozark origin. South of this the Mississippi swings to the west away from the eastern bluff and outlines the Yazoo Basin. Bordering the present alluvial plain is a flight of four river terraces at 350, 200, 100 and 40 feet above the present flood plain. They are separated by low bluffs and mark successive levels of the flood plain during Pleistocene times. Places where the river is undercutting the terraces, e.g. Memphis, Vicksburg and Baton Rouge, are favoured settlement sites.

The flood plain is extended seawards in the delta. The river loses its meandering character and terminates in a crow's-foot pattern of distributaries. Levées, cheniers, back-swamps, bayous, flood-plain splays and extensive but shallow lakes, such as Pontchartrain, diversify the surface. Minor forms of economic importance are the small salt islands rising to heights of as much as 100 feet—the tops of salt plugs often flanked by oil pools.

West of the Mississippi the coastal plain resembles the East Gulf Coastal Plain in its cuesta landscape. The rocks dip south-eastwards under the gulf but there is no equivalent of the Appalachian old land. In the north the Ouachita Mountains fill this role. The western boundary is the Balcones Escarpment, which marks the faulted edge of the uplifted, near-horizontally bedded rocks of the Edwards Plateau. The Black Prairie of Texas is the geomorphological twin of the Black Belt of Alabama; sandy cuestas, clothed with pine, alternate with concentric lowland belts and although the names may differ, the geography either side of the Mississippi is very similar. The Sabine, Trinity and Rio Grande are the principal consequent rivers. The coastlands comprise a long sand-reef broken by a few natural channels. Behind is an extensive lagoon system continued in the drowned lower ends of the major valleys. Unconformities and slight flexures in the rocks of the coastal plain form traps in which petroleum and natural gas have accumulated, making the Gulf coast oilfield one of the most important in the U.S.A. Salt domes and sulphur beds are other economically valuable elements in the rock-pile.

The Great Plains (Region 5a)

East of the Rocky orogenic belt is a long synclinal trough with its axis parallel to the Rockies. Into it rocks of Palaeozoic and Mesozoic ages dip very gently from the east and rather more steeply from the Rockies in the west. They comprise sandstones, limestones, shale and coals. Capping them are sands, silts and gravels spread out in torrential fans by eastward-flowing streams from the Rockies in Tertiary times. Diversity within the Great Plains is determined by the degree of dissection of these rocks, by the presence of localized folding and by associated volcanic intrusions and extrusions. The heart of the region is the High Plains where the extensive gravel spreads are dissected by few rivers. Only the Platte, Arkansas and Canadian rivers rise in the Rockies and flow clear across the plains; but there are others, intermittent in character, which rise on the plains. In the incised valleys the wide, braided river channels are bordered by bluffs and terraces. The Nebraskan plains are covered by extensive eolian sand deposits derived, it is thought, from dry, braided stream channels during the Pleistocene. The Llano Estacado of Texas and New Mexico are the best preserved of the plains, for they have been little dissected, perhaps because of the semi-arid climate.

Surrounding this core of the Great Plains are a number of peripheral and usually lower regions. North of the Pine Ridge scarp the Missouri plateau is more dissected, the valley bluffs are the sites of true badland landscapes and the broad benches are capped by mesas and buttes. These are the product of erosion by the Yellowstone, Little Missouri, Cheyenne and White rivers, all tributaries of the Missouri. In the heart of this semi-arid grassland region is the forested, eroded dome of the Black Hills in which a granite mountain core, rising over 4,000 feet above the plain, is surrounded by concentric hogsback sandstone ridges. Gold, tin and lead ores and scenic attraction are the bases of the economy of this distinctive region.

Dissection also characterizes the eastern border of the plains. The thinning Tertiary gravel cap has been largely stripped away and the rocks beneath carved into east-facing scarps such as the Blue and Smoky hills and vales of Kansas. This type of terrain continues across Oklahoma into Texas in the Osage region (4a). This is the only unglaciated part of the Interior Lowlands, and its scarps are developed on Palaeozoic rocks dipping westwards off the Ozark uplift. In this region are two upfolds on which are developed the Arbuckle and Wichita mountains.

In the west the Tertiary gravel cover has been completely removed and the upturned edges of the older sedimentary rocks, where they dip steeply off the crystalline rocks of the Rocky Mountains, have been eroded into a number of hogsback ridges. The most prominent is formed by the Dakota sandstone. Sink-holes up to a mile wide, some containing lakes, dot the limestone outcrops. The Colorado piedmont is lower than the High Plains, from which it is separated by a pronounced west-facing scarp. In the Cretaceous rocks of this region lignite beds occur; while the extensive lava flows and volcanoes of the Raton section have been eroded into lava mesas, plugs and radiating dyke systems, making this the most diverse part of the plains. The Pecos trough at the south-western margin of the plains, cut below the level of the Llano Estacado by the Pecos River, has potash deposits and the Carlsbad limestone caverns. The Pecos Canyon, cut into the limestone of the Edwards Plateau, links the trough with the Rio Grande. Unlike the Llano Estacado of the high plains, which they adjoin, the Edwards and Stockton plateaux do not have a gravel cover; at 2,500 feet they are lower in elevation by some 1,000 to 1,500 feet. The Central Mineral Region north-east of the Edwards Plateau is yet another eroded dome in the core of which the limestones have been removed and the granite basement exposed. Around this core dissected sheets of limestone form the Lampsas Cut Plains. The relative relief is about 500 feet.

Rocky Mountains (Region 7)

The western third of the United States comprises part of the North American Cordillera. This may be divided into the Pacific and Rocky mountain systems, west and east respectively, of the intervening intermontane plateaux. The Rockies reflect long-continued erosion of a series of parallel upfolds and thrust slices by water and ice. The principal earth movements occurred at the close of Cretaceous times, in the Laramide orogeny. That there have been later movements is clear from the occurrence of at least two uplifted, warped, dissected planation surfaces. The

I (a) EIGHTEENTH-CENTURY PIONEER FARMSTEADS

This picture is a composite study, based upon the observations of Patrick Campbell in the St. John River valley of New Brunswick. It is specifically concerned with illustrating the fence types seen in north-eastern America—the log fence, worm fence, post and rail fence and Virginia rail fence. Behind the cut-over land are represented wooden dwellings and outhouses typical of the day (Source: P. Campbell, *Travels . . . in North America*, Edinburgh, 1793)

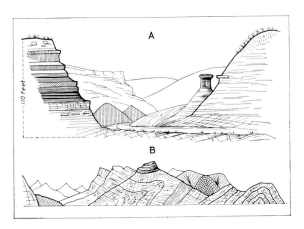

I (b) NINETEENTH-CENTURY FIELD SKETCHES

Field sketches of (A) the Valley of Souri River at La Roche Percé, showing lignite strata, and (B) Athabasca River, first range, from *The Journals relative to the Exploration by Captain Palliser . . . 1857-60*, H.M.S.O., London, 1863. The sketches show the progress in the representation of the countryside seventy years after P. Campbell

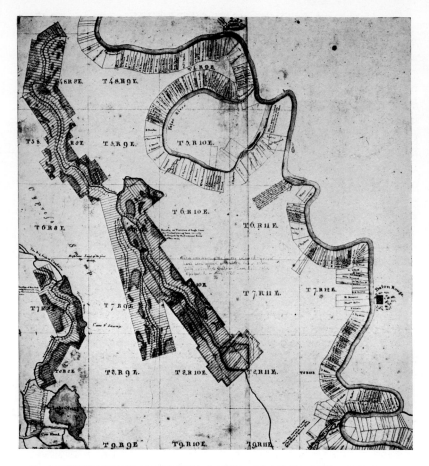

II LOUISIANAN PRIVATE LAND CLAIMS AND SURVEY SUBDIVISIONS

This map, probably drawn in the 1820s, illustrates the type of archive available for studying the expansion of settlement in the lower Mississippi valley. It is in Louisiana south of the base line bordering on the Atchafalaya bay and river, eastward to Bayou Lafourche and the Mississippi river and is accordingly from the same tract as Plate VI

mountains are therefore as much the product of down-cutting by rivers as of orogenic uplift.

The Front Range of the Southern Rockies (7a) reaches altitudes of over 12,000 feet. Long's Peak, the highest, is 14,255 feet. The relative relief is less than this as the mountains rise out of the Colorado piedmont which is itself over 5,000 feet above sea level. Nevertheless, the abruptness of the rise is impressive. Traced northwards, the range bifurcates in the Laramie and Medicine Bow ranges to enclose the Laramie Basin, one of several in the Rockies which were basins of deposition during the erosion of the surrounding ranges. The deposits they contain are akin to the gravels which cover the High Plains east of the Front Range. The Park Range runs parallel to the Front Range and linking them is the cross range of the Rabbits' Ears Mountains, which separates two further basins, North Park and South Park. The word 'park' is an accurate description of the vegetation of these rain-shadow pockets within the Rockies. Their grassland with scattered trees contrasts with the forested lower and middle slopes of the mountains, the windward western sides of which are wetter and greener than the east-facing slopes. The highest levels have been mountain glaciated. Cirques, arrêtes, hanging valleys, lateral and terminal moraines characterize the ground above approximately 8,000 feet. The Colorado River rises on the Front Range and flows westwards, and only a short distance away the South Platte rises in South Park and proceeds to cut its way eastwards across the Front Range. This feat is also performed by the Arkansas River in its Royal Gorge section. Other ranges in the Southern Rockies are the Sangre de Cristo, the Sawatch, the highly dissected lava flows and structural dome of the San Juan Mountains and the less dissected dome of the White River plateau. Where the basement rocks are exposed in the Front Range, the Sawatch range and the San Juan Mountains, gold, silver, lead and zinc are still worked.

The heart of the middle Rockies is not a mountain range but the Wyoming Basin (7b). This is a general term for a number of synclinal basins separated by interior, low, ranges and spurs of peripheral, higher, mountain systems. The Wind River anticlinal range is the largest of the internal ribs. The bordering range in the south is the Uinta anticlinal arch. Corrie glaciers have eaten into the quartzites of its plateau summit forming a distinctive biscuit-board relief, while to the north and south are bold, flanking hogsback ridges. In the west are a series of block-faulted ranges extending from the Wasatch, which overlooks the Great Salt Lake Basin, north to the Grant Tetons, perhaps the most spectacular of all mountains in the U.S.A. In the north-west, the Big Horn Mountains

D

curve round into the Owl Creek range and the Absaroka range, almost enclosing the synclinal Big Horn Basin. The Big Horn range is yet another broad arch breached down to the crystalline basement with marginal hogsback and broader cuesta ridges. Its high summits, over 9,000 feet, have been glaciated but extensive plateau remnants remain. The major rivers draining the Wyoming Basin are the Green, the North Platte and the Big Horn. All three have cut remarkable canyons through the peripheral ranges: the Green its Landore Canyon through the Uintas, the North Platte through the Laramie Range, the Big Horn through both the Owl Creek and Big Horn ranges. There are compelling reasons to suppose that these rivers were superimposed from a cover of Tertiary rocks across the anticlinal structures of these mountains and that they are not antecedent to the folding. Mean annual rainfall in the basin is low, no more than 10 inches. Much is desert and badland erosion forms are common.

Colorado Plateau (Region 8a)

As the folded Appalachians give place westwards to the little-folded rocks of the Appalachian plateau, so the Rockies are followed in the same direction by the plateau lands of Colorado. These are generally not much lower in elevation than the mountains themselves, so that heights of 9,000 feet are not uncommon. The horizontally bedded, many-coloured rocks are a geological column written on the ground. In sequence above the black pre-Cambrian schists are the brown sandstones and blue shales of the Lower Palaeozoic, white Carboniferous limestones, red shales, cream sandstone and grey limestones of the Permian, Triassic chocolate shales, Jurassic red and white sandstones, Cretaceous and Tertiary marls and limestones in shades of pink and white. Add to these the black of the lava flows and one aspect of the beauty of these empty lands can be understood. The major monoclinal flexures and faults lift and drop this thick pile of sediments in a pattern of north-south plateaux and intervening trough-like depressions. Flexuring has produced some intrusive volcanic activity, usually in the form of laccoliths, and in places considerable lava flows have spread over the surface. Cutting across this geological paradise is the Colorado and its tributaries, which have carved breath-taking canyons. The Grand Canyon is a mile deep. The eroded edges of the more resistant beds form scarp after scarp. The Vermilion, White, Echo and Roan cliffs bound plateaux with names such as Kanab, Kaibab, Shivwits, Paunsagunt, Kaiparowits and Tavaputs, and mesas such as Mesa Verde or Black Mesa with its lava cap. Plateau, cliff and canyon developed in multicoloured rocks present a scene little hidden by vegetation in these semi-arid lands unless a plateau be high enough to catch the elusive rain. Small wonder that the manufacturers of colour films issue guides giving the correct exposure at various times of the day and year for many of the well-photographed views across these lands of the Navajo and Pueblo Indian.

Columbia-Snake Plateau (Region 8c)
Horizontally bedded rocks also characterize the extensive plateau of the Columbia and Snake river basins. Lava flows of several ages have spread out layer upon layer until they have almost completely submerged the pre-existing landforms. The region as a whole is shaped like a shallow saucer, the centre of which has been occupied from time to time by lakes so that the lavas are inter-bedded with lacustrine sediments. Large quantities of ground water are stored in the vesicular lava.

Several distinct landscapes are recognizable but their boundaries are far from clear. In the north and west is the Walla Walla plateau drained by the Columbia River flowing in its deep, winding canyon. In the loop of the Columbia the surface of the lava is furrowed by abandoned melt-water channels cut when the Columbia system was the principal outlet of the Cordilleran ice sheet. The largest of these is the Grand Coulee. Set between these channelled scablands and the Rockies is the Palouse region where the soils are developed on a thick cover of loess blown from the glacial outwash deposits. The same type of terrain continues south of the Columbia to the limits of the plateau. Warping and faulting have produced a much more accidented relief west and north of the Columbia.

The Walla Walla plateau is limited southwards by the 10,000-foot-high Blue Mountains, an outlier of the Rockies. The Snake River crosses this range in the sombre Seven Devils Canyon, which for size, if not variety of form and colour, rivals the Grand Canyon. The Snake Basin above the canyon is divisible into two units: the Payette section in which the lavas are buried by lake deposits, and the terraced valleys of the dissecting rivers, reaching depths of 1,000 feet. Upstream is the young lava plateau, sloping up from 2,000 feet in the west to 6,000 feet in the Yellowstone plateau, where it merges with the Rockies. The surface of the flows is little weathered and recent craters are numerous. The Snake itself runs in a narrow canyon, its long profile broken by a number of waterfalls where it cuts through the successive lava flows. In the south-west the lava plain forms the Great Sandy Desert, an area of inland drainage transitional to the Basin and Range region.

Basin and Range Province (Region 8b)

This vast region is a symphony of sight, but little sound, in which the themes are dissected block-fault mountains, mountain-foot pediments, bajada fans and playa salt flats. In the northern half, or Great Basin, such drainage as there is in these desert climes is completely internal. In the south the Colorado and the Rio Grande rivers reach the coast so that their basins are open.

Carson Sink, in the Great Basin, is today the graveyard of the Humboldt and Carson rivers; but in Pleistocene times it was the centre of a much larger Lake Lahontan. The basin then experienced pluvial periods when rainfall was high, whilst farther north and on the adjacent mountains ice caps held sway. The present Great Salt Lake and the Bonneville Salt Flats are the successors of glacial Lake Bonneville, which attained a depth of 900 feet and overflowed northwards into the Snake Basin. Unlike Lake Lahontan, which had no outlet, it was therefore a fresh-water body. The shorelines of these ancient lakes were cut on the flanks of the bounding mountains and on those which rose above their waters as islands. From a distance they have the appearance of contours etched on the mountain sides.

The mountains rise some 4,000 feet above the floors of the bolsons and may reach heights of 10,000 feet above sea level, but collectively they do not occupy as great an area as the intervening basins. Characteristically, they trend north–south and are rarely more than 50 miles long, and their steep edges are straight because fault-guided. Dissection by hundreds of ephemeral streams gives them a rugged appearance in detail. Mineral ores are common, e.g. copper ores quarried and smelted close by the Great Salt Lake west of Salt Lake City (Figure 48). The salt deposits are also exploited.

In the Californian portion of the Great Basin is Death Valley, a basin the floor of which is 276 feet below sea level, climatically the hottest spot in North America. The Mojave Desert is continued southwards in the Sonoran Desert. The Salton depression at the head of the Gulf of California is another type of enclosed depression. The Colorado River has, in recent times, built its delta across the head of the gulf, the lower part of the basin enclosed being over 300 feet below sea level. It includes the irrigated lands of Imperial Valley and the relic Salton Sea.

Sierra Nevada and Cascade Ranges (Region 9a)

The western margin of the Great Basin is the impressive fault scarp front of the Sierra Nevada. This range is essentially a block of granites 400 miles long and 80 miles wide, which has been tilted westwards so that its eastern edge and crest line is raised some 3,000 feet above the basin, whilst its back slope descends from heights of over 10,000 feet to the Central Valley. As a result of successive rejuvenations a series of parallel streams draining westwards have cut deep troughs in the peneplained upland with its granite domes. The Feather, Kings, American and Merced troughs were further deepened by glaciers which descended from an ice cap on the high Sierra. The most accessible of these valleys is that of the Merced—alternatively named Yosemite. The granites are batholithic intrusions into much-folded Palaeozoic and Mesozoic rocks which contain the mother lodes of gold. It was the discovery of gold-bearing river gravels in old stream channels derived from these lodes which led to the gold rush of 1849 (visits to their camps prompted R. L. Stevenson to write the *Silverado Squatters*). The white sheen which the Sierra granites impart to the landscape is accentuated by the snows of winter and diversified by the greens of forests which contain groves of *Sequoia sempervirens*.

The Cascade Range is in general more of a horst than a tilted block like the Sierra Nevada. Topping its granite base are a collection of volcanic cones. Lassen Peak, the southernmost, is a plug volcano last active in 1922. In the same area are the eroded stumps of a much larger volcano and Mount Shasta, a beautifully symmetrical cone. North of these the best-known features are Crater Lake, Mount Hood and Mount Rainier. In the central Cascades, broad upwarping has perhaps been matched by the downward erosion of the Columbia River, which therefore crosses the Cascades in an antecedent gorge.

Between the ranges just described and the Coast ranges is a discontinuous trough comprising a number of basins (8b). The best defined is the Central Valley of California between the Sierra Nevada and the coast ranges. Its near-flat floor is made up of a number of coalescent, low-angle alluvial cones, consisting for the most part of detritus from the erosion of the Sierra. The longitudinal rivers draining the valley—the Sacramento from the north, the San Joaquin from the south—have been pushed to the west side by these cones as relatively little material was eroded from the semi-arid, streamless east slope of the coast ranges. In the still more arid south the Tulare Basin is cut off from the San Joaquin.

The mirror image of the California Valley is the Willamette-Cowlitz-Puget Sound lowland in Oregon and Washington. The 150-mile-long Willamette Valley lies between the Cascades and the Coast ranges. Its alluvium-filled floor is threaded by the braided Willamette and crossed by the discordant Columbia River. North of the Columbia the depression is floored by glacial deposits from the Cordilleran ice sheet, the terminal moraine of which is a few miles south of the head of Puget Sound. South of the morainic hill belt there are vast spreads of outwash sands and gravel extending to the Columbia, whose valley was the outlet for the melt-water. The intricate shoreline pattern of Puget Sound is a result of the partial drowning of a rolling morainic topography.

The Los Angeles Basin is also, in some senses, a mirror image of the Central Valley, for its western half is beneath the sea. The only remains of the western wall are the islands of Santa Catalina and San Clemente. As well as the Los Angeles lowland, filled with alluvial fans and open to the sea, there are a number of closed basins and minor mountain ranges. Across the fans flow ephemeral streams which break up into a number of distributaries before their water sinks into the porous subsoil to emerge down-slope in springs. Coastal plateaux extending back from the cliff tops are clear evidence of former higher sea levels. The flat-floored valleys are frequently blocked at their seaward end by bars through which the waters of the lagoons seep into the sea (Figure 46).

The Coast Ranges (Region 9c)

The westernmost mountains in the United States are as diverse in character as any of the others. In the south the Californian ranges are built of folded and faulted sediments dissected by rain and rivers into a multitude of ranges. They are oriented at an angle to the coastline so that the frequently fault-guided valleys, such as the Salinas, open out north-north-westwards to the Pacific or, in the case of the Santa Clara Valley, to San Francisco Bay—an arm of the Pacific reaching through the Golden Gate. Tributary to the bay from the north is the vine-filled Napa Valley. The ranges rarely reach 4,000 feet, their burnt, brown summer appearance changing north of the Golden Gate into one of green redwood forest. The valley pattern reflects that of the trellised drainage which is more adapted to faults than to the strike outcrops of weaker rocks. The major line of weakness is the San Andreas fault: movement along this fault is the common origin of earthquakes, such as that of 1906 which, together with subsequent fires, destroyed most of San Francisco. There is no true coastal plain, although elevated wave-cut benches up to heights of 1,000 feet border the coast. The last movement of sea level has been a positive one and drowning has created Tomales, San Francisco and Monterey bays. The east-west San Rafael and Santa Ynez ranges, part of the Los Angeles system, form a link with the Nevada fault-block system.

In northern California and southern Oregon the Klamath Mountains are a much dissected, uplifted peneplain with monadnocks cut in igneous rocks. The summits reach to 8,000 feet, whilst the general upland level is 4,000–6,000 feet. The Trinity, Klamath and Rogue rivers cut across the whole system to the Pacific. The summit level in the Oregon and southern Washington sections is less than 2,000 feet where gently folded sediments in a north-south oriented anticlinorium have been dissected by the transverse Umpqua, Columbia and Chehalis rivers and their tributaries. The Olympic Mountains are a dissected upland reaching 5,000 feet with high peaks up to almost 9,000 feet. From Juan de Fuca Strait south to Cape Mendocino the coastal plain is formed principally by wave-cut benches on the seaward side of the ranges.

2. THE GLACIATION OF NORTH AMERICA

At some time during the Pleistocene, ice covered the continent southwards to a line approximately defined on Figure 7. It runs through Long Island, north-west across the Appalachians almost to Lake Erie, then follows approximately the Ohio Valley to its confluence with the Mississippi, and so up the valley of the Missouri to the Rockies and across the Cordillera along latitude 48°N to the Pacific coast.

There were two major ice caps. The lesser consisted of ice born on the Rockies and Coastal Mountains of British Columbia. It was nurtured from youthful piedmont glaciers to a mature ice sheet in the plateau between those ranges until it was sufficiently grown to shed its ice seawards to the Pacific and landwards between the nunatak peaks of the Rockies to the Alberta step of the Prairies. Here the Cordilleran ice met and held back the much larger, westward-moving Laurentide ice sheet. The line of contact must have varied with the relative strength of the two sheets, but the evidence of erratics suggests that the mountain ice was at times able to penetrate up to 50 miles east of the Rockies.

The stages in the growth of the Laurentide ice sheet are a matter of conjecture. Suffice it to say that the high eastern rim of the Laurentian Shield must have cradled its early stages, and that it was joined later by ice from other centres. It seems likely that there were ultimately at least two centres of ice dispersion, one in Labrador and a second in Keewatin.

Within this area affected by the two ice sheets the Wisconsin driftless region was neither ever covered by ice nor at any single time completely surrounded by ice. Beyond the general southerly limit subsidiary ice caps existed on major mountain masses such as the Northern and Colorado

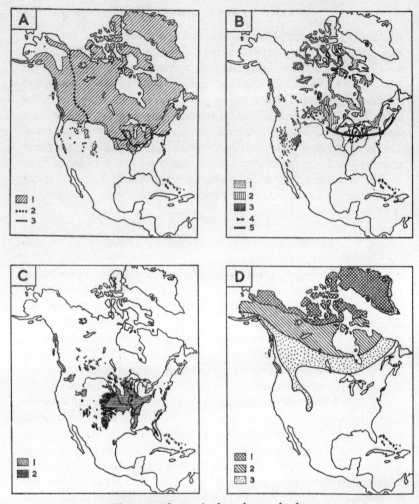

Figure 7 *The Work of Ice, direct and indirect*

A. *Extent of Pleistocene Glaciation*
   1 Glaciated Area. 2 Ice Divide. 3 Limit of Wisconsin Glaciation
B. *Related Water Features*
   1 Areas of post-glacial Marine Invasion. 2 Pro-glacial Lakes. 3 Lake Bonneville and associated pluvial Lakes. 4 Major overflow Channels. 5 Southern Limit of crustal Depression
C. *Associated Wind-borne Deposits*
   1 Loess. 2 Sand
D. *Present State of Permafrost*
   1 Continuous. 2 Discontinuous. 3 Sporadic
   (with modifications, from R. F. Flint, *Glacial and Pleistocene Geology*, New York, 1957)

Rockies and the Sierra Nevada-Cascades. There were also many hundreds of valley glaciers on less extensive ranges.

The Laurentide ice advanced into a broad lowland between the Appalachian plateau and the High Plains, penetrating farthest where resistance was least. The great southward bulge of the ice front to the head of the Mississippi embayment was one result. In detail the ice front was lobate, each lobe conforming to a major valley lowland or basin. This is clearly shown in the festoon pattern of the moraines across the states of Indiana, Illinois and Iowa.

The ice has left a legacy of both erosional and depositional landforms. In general terms the hard rock core of the Canadian Shield was the area of erosion, whilst the till plains of the Middle West to the south and the Prairies to the west were at the receiving end. There is some truth, therefore, in the contention that, in the glacial equation, Canada's loss has been the United States' gain. But, as aerial photography has revealed in the last 15 years, there are plenty of depositional forms, particularly eskers, on the Shield.

The pre-glacial drainage pattern has been greatly modified if not completely destroyed wherever the ice penetrated. Beneath the drift there are buried pre-glacial valleys across which the present river valleys cut without regard. Along the edge of the ice sheets ice-marginal melt-water channels eventually became linked to form regular river valleys. In this way the Ohio and Missouri rivers were formed.

Along the retreating edge of the ice sheets, where the land surface sloped back under the ice, a series of pro-glacial lakes emerged. The present chain of lakes around the margin of the Laurentian Shield, including the Great Lakes, is the last vestige of this phase. Their geographical significance also rests in the channels cut by their outlet streams, which provide easy access from one river system to another. An example is provided by that between Lake Michigan and the Mississippi via the Chicago and Des Plaines rivers, a route followed by melt-water from an ancestor of Lake Michigan, called Lake Chicago. Flat, silt and clay-covered old lake floors, with beach ridge and delta-etched margins, are an important element in the physical landscape. These are all that remain of glacial lakes Agassiz (the Red River lowlands), Barlow-Ojibway (the Clay Belt) and the succession of lakes, Maumee, Chicago, Saginaw, Whittlesey, Warren, Duluth, Iroquois, Algonquin and Nipissing in the Great Lakes Basin.

As the ice margin retreated so new outlets were opened, old ones abandoned, and the lake pattern changed. A complicating aftermath of deglaciation has been the gradual rebound of the earth's crust after its depression under the load of the ice sheets. This has been sufficiently critical at times to tip water back through an abandoned routeway. In this manner, the North Bay outlet of Lakes Huron, Superior and Michigan into the Ottawa River was abandoned in favour of a former outlet, still used, via the St. Clair River into Lake Erie. The isostatic recovery of the land lagged behind the melting of the ice, and the sea was able to penetrate up the St. Lawrence, and across the Hudson lowlands, leaving evidence in beach ridges, wave-etched benches and a superficial covering of marine sediments. These are now high and dry after further recovery of the land, a process still continuing north of the hinge line at rates up to 50 cm. per century.

In Pleistocene times vast expanses of outwash deposits, tracts of newly exposed till and desert lands formed sources of much wind-born material. This covers extensive areas in the United States (cf. Figure 7) in the form of sand dunes, born of the coarser materials and loess, of the fine material. These are yet another group in the kaleidoscope of soil parent materials characteristic of the ice-influenced lands of North America.

At present much of northern Canada has a perennially frozen subsoil or permafrost, although there is little ice on the surface. Associated phenomena are summed up in the term peri-glacial, including frost shattering, solifluction and patterned ground. Peri-glacial influences were felt much farther south at the time of maximum ice advance and they remain as relic features in the landscape.

The sequence of events during the Pleistocene was complex, but the following stages are generally recognized:

Glaciation	*Interglacial*
WISCONSIN	
	Sangamon
ILLINOIAN	
	Yarmouth
KANSAN	
	Aftonian
NEBRASKAN	

The Wisconsin is overwhelmingly important in that most of the moraines, till plains, drumlins, eskers, lake deposits and loess are of this age. The deposits of the older glaciations, particularly the Nebraskan and Kansan, have little influence upon the present land surface. The Wisconsin drifts are divisible into four stages: Mankato or Valders, Cary, Tazewell and Iowan. The limits of each are marked by terminal moraines, e.g. the Valparaiso moraine encompassing the southern edge of Lake Michigan is the end moraine of the Cary stage.

3. A CROSS-SECTION OF THE GLACIATED LANDS

The New England-Acadian region (Region 2e)

The New England portion of the crystalline Appalachians pokes two fingers south-westwards. On the extremity of the southern finger stands Manhattan, one of the five component boroughs of New York City. The other, the Reading prong, reaches out towards the Blue Ridge.

The most extensive of the physical subdivisions of New England is the upland which runs from Connecticut to New Brunswick. It is a peneplain uplifted to heights of 2,000 feet at its inner margin and sloping seawards to less than 1,000 feet. At many points it is overlooked by isolated mountains, the type example of which is Mount Monadnock. Everywhere the work of ice is ubiquitous in lake basins, bare rock surfaces, eskers, outwash deposits, waterfalls and erratics. The second subdivision, the coastal lowland, is broken by arms of the sea and fringed by thousands of islands—a shoreline which was the inspiration of D. W. Johnson's pioneer study of coastal landforms. The glacial deposits include two famous drumlins: Bunker Hill outside and Knob Hill inside the city of Boston. Elevated wave-cut benches and beach lines testify to a complex history of glaciation, marine invasion and crustal uplift.

The White Mountains culminate in the corrie glaciated Presidential Range. The weather observatory on the summit of Mount Washington stands at 6,920 feet, 1,000 feet above the tree line. The range continues north-eastwards in the Megantic Mountains to Mount Katahdin, an isolated table mountain rising 4,000 feet above the lake-pocked New England upland. Vermont, 'the green mountain state', takes its name from the Green Mountains which parallel the White Mountains west of the Connecticut Valley. Their granites and marbles (and asbestos in their Canadian continuation) reflect the crystalline character of the rocks. The Taconic Mountains west of the limestones and slates of the Berkshire lowland are lower still; their 2,000-foot summits look westward over a low plateau to the Hudson Valley.

Between the White and Green mountains is the Connecticut Valley, a broad lowland eroded on a down-faulted outcrop of Triassic sediments softer than the bounding crystalline rocks and overlooked by angular trap ridges. During the decay of the ice in New England the valley was occupied by glacial Lake Hitchcock, the floor of which, now terraced by the incision of the Connecticut River, is a most favoured site for settlement and cultivation. The lower Connecticut, turning its back upon the direct southern route to the sea and cutting obliquely across the crystalline basement to the coast, provides one of the earliest-described examples of superimposed drainage.

The parallel mountain ranges are continued north-eastwards into the Eastern Townships of Quebec. Long lowlands, sometimes lake-filled, separate mountain ranges which merge northwards into the broad, flat-topped, treeless Shickshock mountains, standing 4,000 feet above the estuary of the St. Lawrence. They terminate in the high cliffs of the Gaspé peninsula.

The form of the New England upland is repeated across the Canadian border. Broad glaciated upland plateaux on the resistant rocks alternate with lowlands on the weaker sedimentary strata. In Nova Scotia the upland averages some 600 feet above sea level, but its glacially scoured surface slopes south and east to the Atlantic and Gulf of Maine. On its western side the deep and narrow Annapolis-Cornwallis Valley marks the edge of the northernmost down-faulted Triassic outcrop, the outline of which is neatly described by the shores of the Bay of Fundy. North Mountain, the west side of the Annapolis Valley, is a 120-mile-long, flat-topped, 550-foot-high outcrop of trap rock. Flat or gently dipping Carboniferous rocks underlie the Cumberland-Pictou lowland, the Eastern Plain of New Brunswick and Prince Edward Island.

The Laurentian Shield (Region 6)

The highly contorted Pre-Cambrian rocks of the Shield must have been uplifted and worn down many times. The last peneplain was raised, warped, faulted, incised by rivers and subject to a succession of continental glaciations to shape the present surface of the Shield. Its monotonously rolling skyline is only occasionally broken by an upstanding residual summit; but in detail ice-scoured basins alternate with broad whale-backed, ice-smoothed ridges and snaking eskers. Water, muskeg, trees and tundra cover its surface as slope, soil and climate permit. Some 25 to 35 per cent of the surface is lake-covered.

The Shield is a saucer, the raised rim of which reaches 5,000–6,000 feet in the Tornagat Mountains of Labrador, 2,000 feet above the Strait of Belle Isle, and declines to 1,500–1,000 feet south of Quebec City. Ice pouring over the rim from interior Labrador-Quebec has carved spectacular fiord country. Breaks and sags in the rim occur at Ungava Bay, Hamilton Inlet, north of the Ottawa River (where a broad col gives access to the clay belt) and in the Nelson col of Manitoba. Through the Nelson col the waters of the Prairies drain via Lake Winnipeg and the Nelson River from beyond the rim to the centre of the saucer in Hudson Bay. Overlying the basement rocks on the southern shore of Hudson Bay are virtually unfolded Palaeozoic sedimentary rocks, in turn covered by glacial deposits, drumlins, eskers and beach ridges of a later age. In the Lake St. John lowland—an almost self-contained unit, drained by the Saguenay to the St. Lawrence—similar rocks are covered by lacustrine and morainic deposits. Spanning the provincial border between Ontario and Quebec is a belt of clay deposits laid down in the pro-glacial waters of the ephemeral lakes Barlow and Ojibway (Figure 55).

There are two extensions of the Shield into the U.S.A. In the west are the iron-bearing rocks of the Superior upland west and south of Lake Superior. In the east are the much higher Adirondacks. The Adirondacks are linked geologically to the Shield by the Frontenac axis which is crossed by the St. Lawrence River in the shallow Thousand Islands section between Kingston and Montreal, and by the North Bay-Ottawa River channel between Georgian Bay and Montreal. The obstruction of the Frontenac axis was the physical *raison d'être* of the St. Lawrence seaway.

The drainage pattern of the Shield is glacially deranged but certain valid generalizations can be made. In the Labrador peninsula rivers radiate from the height of land along the Quebec border around the iron fields of Schefferville and Burnt Creek. Conversely, Hudson and James bays are the foci of a centripetal pattern in sympathy with the saucer shape of the Shield. In the south-west the influence of the Nelson extends via Lake Winnipeg and the Saskatchewan River beyond the Shield to the foothills of the Rockies. In the north-west low gradients, the multitudes of lakes—and lack of information about them—often make it impossible to say which way many of the rivers flow. The ring of large lakes marking the edge of the Shield—Great Bear, Great Slave, Athabaska, Winnipeg, Superior, Huron and Ontario—are located where ice was able to excavate deep basins in the less resistant sedimentary rocks lapping on to the Shield along its landward margin.

The Interior Lowlands (Region 4)
The geographical heart of the continent lies in the belt of lowland which margins the crystalline shield from the St. Lawrence Valley in the east to

the Manitoba lowlands in the west. The twin themes in its landforms are cuestas and glaciation. The cuestas are reflections of the alternating resistant and less resistant rocks which dip off the Shield. The geographical influence of the Shield is therefore felt far beyond its physical limits.

East of the Mississippi three different combinations of the two controls occur. In the 'lake district' focused on lakes Michigan, Huron, Erie, and Ontario (8c), the cuestas are prominent. The Niagara limestone cuesta forms the axis of the Niagara peninsula and its scarp overlooks Lake Ontario. It underlies the Indian peninsula and Manitoulin Island between Lake Huron and its adjunct Georgian Bay, and in a similar fashion separates Green Bay from Lake Michigan in the Door peninsula. Less prominent cuestas run parallel to it both north and south. The subsequent lowlands between the cuestas have been ice-scoured to form the lake basins of Michigan, Huron, Ontario, Green Bay and Georgian Bay. Their arcuate pattern reflects the belted outcrop of the Palaeozoic rocks around the Michigan structural basin.

The bottoms of the lake basins are 200 to 300 feet below their rims; a fact expressive of the gouging effect of the several ice lobes. Lake Superior, located almost entirely on the Shield, fills a similarly deep basin, but Lake Erie is very much shallower. The water level in Lake Superior is 602 feet, in Lake Huron 581 feet; and between are the rapids at Sault Ste. Marie. Lakes Michigan and Huron are at the same altitude and there are no rapids in the Strait of Mackinac. Between Huron and Erie is a fall of 9 feet along the St. Clair River and through Lake St. Clair (576 feet). The Niagara River between lakes Erie and Ontario drops from 572 to 246 feet, of which approximately half is in the Niagara Falls where the river tumbles over the Niagara limestone scarp into a 9-mile-long gorge cut by headward erosion into the cuesta.

No less prominent are the myriad smaller landforms attributable to the work of ice. Terminal moraines festoon the land and reflect the lobate form of the ice. Where two lobes of ice were in juxtaposition the angle between was filled with interlobate morainic deposits. Today these form strong knots along the strung-out and sometimes tenuous lines of moraine. The morainic hills along the axis of the Ontario peninsula originated as interlobate moraine, between the Erie and Huron ice lobes. Drumlin fields are particularly prominent on the lowlands south of Lake Ontario, in northern New York State and in eastern Wisconsin. Over broad areas around the margins of the lakes, old lake floors and their associated beaches mark the sites of former pro-glacial lakes. The beach ridges and moraines form threads of higher and drier land amongst the wetter mucklands of the old lake floors.

The St. Lawrence Valley lowlands (8b) north-east of the Frontenac axis are underlain by near-horizontal limestones and shales eroded to form a lowland region, now buried by a spread of glacial detritus and the relics of post-glacial submergence by the Champlain and Ottawa seas. The lowland, shaped like a dumb-bell, has a narrow centre section where the faulted margins of the Shield and the Appalachians are separated, at Quebec city, by not much more than the width of the river itself. Upvalley the floor broadens into a plain 70 miles wide and 100 to 300 feet above sea level. Rising above it are the Monteregian Hills, volcanic eminences from 500 to over 1,000 feet high. Mount Royal in Montreal city is one of these. The lowland continues south-eastwards into the Champlain-Hudson Valley. The northern end of the 'dumb-bell' is mostly below sea level in the estuary of the St. Lawrence but a fragment remains as dry land in Anticosti Island.   •

South and west of the Great Lakes, in eastern Ohio, Indiana and Illinois, are the very gently rolling till plains characteristic of the Corn Belt (8d). These vast spreads of ground moraine of Iowan and Tazewell age have little relief apart from lines of low morainic hills marking the limit of the ice sheet at successive stages of advance during the Wisconsin glaciation. Within one square mile, a section on the Land Survey system, altitudes frequently differ by less than 10 feet. These 'upland' areas, with their uncertain drainage, contrast with the major valleys of rivers such as the Wabash and Miami cut between steep bluffs well below them. The valleys carried melt-water from the ice front and were filled with outwash deposits now dissected into flights of terraces. Spreads of sandy outwash in the valley trains and outwash plains were sources of windborne material, so that sand dunes were formed on the eastern, downwind, side of such spreads, while loess was carried farther and deposited against the valley bluff and over the upland beyond.

West of the Mississippi, in southern Iowa and northern Missouri, the till plains are markedly dissected. The drift only remains on the flat ridge tops, valleys are prominent and morainic features non-existent. The area was covered by ice only in Nebraskan and Kansan times, never in the Wisconsin, so that dissection has destroyed the depositional landforms.

In south-western Wisconsin is an area of some 15,000 square miles entirely surrounded by drift-covered land but never itself covered by ice. At the same time this driftless area was never at any stage surrounded by ice. It lay between the Superior and Michigan lobes of successive ice sheets and its scarps, although modified by peri-glacial processes, have never felt the pluck and scrape of ice.

North and west of the driftless area, in Wisconsin, Minnesota and the Dakotas, are more till plains related to the Iowa and Dakota lobes of the late Wisconsin ice (Mankato) age. Knob-and-kettle topography, elongated drumlin forms and terminal morainic loops diversify these otherwise featureless lands. The originally west-to-east drainage was diverted by the ice into the present north-to-south Missouri.

In the U.S.A., bedrock is buried by drift but north of the 49th parallel the Niagara limestone scarp emerges in the Western Lakes region (4c) and forms a feature separating Lake Winnipeg from lakes Manitoba and Winnipegosis. The lake basins are carved in the subsequent lowlands either side of the cuesta and their orientation reflects the strike of the Palaeozoic rocks which here curve around and dip off the western edge of the Canadian Shield. But for the most part the landscape of southern Manitoba and central Saskatchewan is dominated by landforms of glacial deposition. During deglaciation when the ice front, although it had retreated on to the Shield, still blocked the Nelson River outlet of the Prairie drainage to Hudson Bay pro-glacial lakes were formed. The largest was Lake Agassiz, which occupied an area 700 miles long by 250 miles wide and was at its maximum 700 feet deep. The water was ponded up between the ice and the east-facing Manitoba scarp. Arms of the lake extended westwards along the gaps cut by the Saskatchewan and Assiniboine rivers in this scarp and southwards along the Red River Valley. It was in this direction that the waters of the lake overflowed into the Missouri system.

The Mackenzie lowland (4f) is a continuation of the subsequent lowland underlain by gently westward-dipping Palaeozoic rocks, etched by ice and liberally covered by its deposits. The ever-faithful Niagara limestone cuesta borders the Great Slave Lake. In the Franklin Mountains the rocks have been folded into linear ranges which rise above the lowland.

In the Prairie Provinces of Canada between Winnipeg and the Rockies the land rises from about 600 feet to some 4,000–5,000 feet across a widely spaced series of scarps and plateau steps. The lowest tread, formed by the flat floor of former Lake Agassiz, is bordered on the west by the Manitoba scarp, the east-facing edge of the Lower Cretaceous rocks. The scarp is dissected into a series of hill masses, e.g. Duck Mountain, Riding Mountains and the Pembina Hills, which when seen from the Manitoba lowland have a relief of up to 1,800 feet. They rise

III LOWER PIEDMONT OF PENNSYLVANIA

Strip-cropping and contour ploughing give rise to this remarkable landscape design from Lancaster County on the lower piedmont of Pennsylvania. In spite of it some erosion gullies can be seen. Land ownership predates the introduction of township and range surveys. Diversified farming makes this one of America's richest provinces in value of farm output

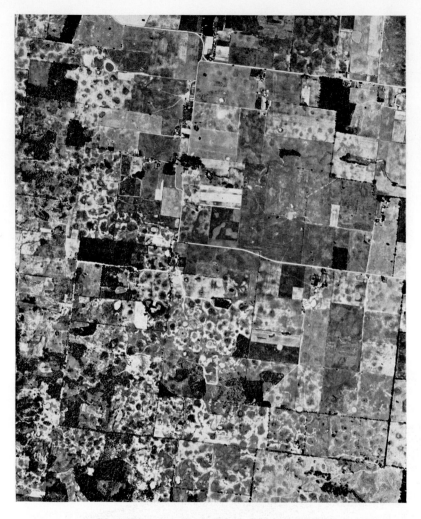

IV KARST LANDS OF KENTUCKY

The patchwork of farmsteads within their framework of townships from Hardin County (Ken.) is much pitted and pocked by sinkholes. The sinkholes impede cultivation and exaggerate tendencies to erosion in a county where burley tobacco is the chief crop

some 500 feet only above the second, or Saskatchewan, step, still part of the Western Lakes region (4e). Deltas deposited in the waters of Lake Agassiz by the Assiniboine and Pembina rivers form gently inclined planes from one step to the next. The Saskatchewan step is on average 2,000 feet above sea level, although in the Touchwood Hills, Moose and Turtle mountains it rises across very gentle slopes to 2,500 feet. It is flattest where formerly pro-glacial lakes existed. Elsewhere, drumlin ridges, morainic hills, shallow lakes and abandoned melt-water channels diversify the surface configuration. The flat floors of the Qu'Appelle and Souris rivers run between steep valleyside bluffs, so that to the traveller's eye the valleys are hidden and horizons seem unlimited. The Prairies are an ocean of land whose monotony, like that of the sea, turns attention to the skies above them.

The Glaciated Great Plains (Region 5b)

The glaciated Great Plains of Canada and the northern U.S.A. are reached across the Missouri Coteau, the east-facing scarp of the Upper Cretaceous rocks. West of this the third, or Alberta, step climbs gently from 2,500 to 4,000 feet. Above it rise mesas of Tertiary rocks which form such hilly areas as The Old Man on His Back Plateau and the Cypress Hills, below it are incised the Saskatchewan River and its tributaries. Gentle flexures in the coal-bearing Cretaceous rocks are the sites of oil and gas pools. The southern limit of glaciation is approximately indicated by the line of the Missouri River, which came into being as an ice-margin stream. In the north the Peace River plains, cut across Lower Cretaceous rocks, are drained northwards by the deep-cut valleys of the Athabasca, Peace and Liard rivers. Muskeg and swamp characterize the valley floors, forests cover the drier uplands.

The Northern and Alaskan Rockies (Regions 7c, 7d)

The Rocky Mountains thrust their summits 5,000 feet above the Prairies to altitudes of over 10,000 feet. They are one of the most typically Alpine mountain chains in the world. The greys of their limestones, the winter whites of ice and snow, the summer blues of sky and lake and the year-round greens of the forest combine seasonally in travel-poster brilliance. They are carved by water and ice from great fault-bound blocks which have been folded and thrust eastwards. They are over 1,000 miles long, but no more than 80 miles wide. Only the Peace River crosses them completely but the passes in the south—Crow's Nest, Yellowhead and Kicking Horse—are the more important.

E

Beyond them is the Rocky Mountain trench, a thrust-fault guided slit between 2 and 10 miles wide, occupied from north to south by the Liard, Peace, Frazer, Columbia and Kootenay rivers. Its terraced floor is 2,000–3,000 feet above sea level. Traced southwards into the U.S.A. it forms the western limit of the Montana Rockies where the rocks have also been pushed eastwards in slices over thrust planes such as the Lewis Thrust. The structural ribs of this complex system have been laid bare by mountain glaciation in such scenically attractive areas as Glacier National Park. Igneous intrusions in the vicinity of Helena, Butte and Anaconda produced major deposits of non-ferrous metals, principally copper ores. West of the trench are the fault-block Bitter Root ranges, with intrusions of igneous rocks in the Cœur d'Alene lead-mining area. Still farther west are the granite-based Salmon River Mountains, their plateau summits deeply entrenched by the Salmon and its tributaries and their southern margin lapped around by the lavas of the Snake Basin.

In Canada the Columbia system of mountains between the trench and the plateau comprises the Purcells, Selkirks, Monashee and Caribou ranges. The quartzite horst of the Purcells is separated from the higher Selkirks by the Purcell trench, the floor of which is drained by the middle Kootenay. The 8,000-foot summits of the former were over-ridden by the Cordilleran ice sheet; the 11,000-foot summits of the latter always stood above it. The Selkirk or Columbia trench separates the Selkirks from the Monashee Mountains. These 6,000–8,000-foot-high mountains, because of their north-south trend, are in contact with the northern end of the Cascades at their southern extremity. Between the two the intermontane plateaux are pinched out. North of the Monashee range the rounded and forested summits of the Caribou Mountains stand 6,000 feet above sea level inside the great bend of the Frazer River.

The British Columbian Rockies end at the Heard River crossing. At their northern end they broaden westwards in the Cassiar Mountains. The Rocky system continues northward in the Mackenzie, Selwyn and Ogilvie mountains. East of the Mackenzie River are the Franklin Mountains, on whose flanks are fold structures yielding oil at Norman Wells. The Richardson Mountains link the Mackenzie Mountains north of the Peel River crossing with the Brooks Range, the Alaskan termination of the Rockies.

The Canadian Plateau (Region 8d)

The interior basins of the Frazer and Nechaka comprise rolling plateaux separated by deeply incised valleys. Altitudes vary from 4,000 to 5,000 feet in the south to 3,000 feet near Prince George. As the Frazer flows southwards the degree of incision increases correspondingly. Glaciation in this semi-arid region was by ice which flowed from the surrounding ranges and collected on the plateaux. The uplands continue northwards in the more rugged Stikine plateau between the Cassiar and Coast mountains, and drain by way of the Stikine and Taku rivers directly to the Pacific. Farther north again is the much larger Yukon plateau—for the most part unglaciated because of lack of precipitation. It is bordered on its eastern side by the Pelly Mountains, which in turn overlook the northward extension of the Rocky Mountain trench.

Central Alaska (Region 8e)

The northern end of the Yukon plateau extends into Alaska, but the Alaskan intermontane plateau has been dissected by the Yukon and Kuskokwim river system into a land of plains and uplands. The principal relief regions are, from north to south, the Koyukuk upland and Seward peninsula which border the Brooks Range, the Yukon plains, the Kuskokwim plateau and the Kuskokwim plains. Drainage is westwards to the Bering Sea.

The Pacific Mountains in Canada and Alaska (Region 9)

The tripartite division into eastern and western ranges and central trough holds in British Columbia and Alaska as in Washington and Oregon, but submergence is much greater than south of the 49th parallel. As a result, the Coast Range of the U.S.A. becomes the island chain of British Columbia and the drowned trough forms the sea passage between the islands and the mainland. The Coast Mountains of British Columbia are the lineal continuation of the Cascades (not the Coast Ranges) of the U.S.A. They consist largely of intrusive granites. Their summits reach over 13,000 feet and induce heavy precipitation. During the Pleistocene, heavy snowfall produced vast quantities of ice, which streamed inland to feed the Cordilleran ice sheet and seawards to carve valleys into deep troughs which were later drowned to form fiords. Ice caps still remain along the Alaska boundary. The arc of the Alaska Range, rising to Mount McKinley, the highest point in North America (20,300 feet), is in part volcanic. In this it resembles the Cascades rather than the Coast Mountains. At its western extremity the mountain system passes into the volcanic Aleutian Range and ultimately, where it is drowned by the ocean, into the Aleutian chain of islands, also volcanic in origin.

The central trough is represented by the drowned, tortuous, island-studded inland passage between the mainland fiords and the sedimentary or volcanic ranges of the Vancouver, Queen Charlotte, Prince of Wales, Baranof and Chichagof islands. Immediately to the north the ice-covered St. Elias Mountains are welded to the mainland and the central trough is squeezed out. Mount Logan, their highest peak (19,850 feet), is also the highest in Canada. Between the Chugach and Kenai ranges and the Alaskan Range the central trough is once more developed in the form of a number of lowland basins, partly drowned by Cook Inlet and including the Matanuska Valley. The final expression of the western Pacific mountain range is found in the rugged fiord lands of Kodiak Island, while the strait between it and the mainland is a last glimpse of the central trough.

THE RECORD OF THE CLIMATE

To the record of the land must be added the record of its climate. The earliest climatic statistics in America date from the seventeenth century. They are commonly attributed to the Rev. John Campanius at Swedes' Fort, Delaware, in 1664. Instrumental records were made in the 1730's, but continuity in recording waited until the mid-nineteenth century, when both the Smithsonian Institute and the military authorities began to build up their meteorological registers. Lorin Blodget's *Climatology of the U.S.A.* (Philadelphia, 1857) first brought a little order into the chaos of collected data. In 1890 Congress created the Weather Bureau as an office of the Department of Agriculture and its network of observers embraced ten times as many volunteers as professionals down to the second world war.

Canada's earliest series of records are ascribed to Thomas Hutchins, 1772-3, from York observatory of the Hudson's Bay Company. A unified picture from the east became possible with the establishment of the Dominion Meteorological Service in 1871 and the exchange of data with the U.S.A. Regular observations from the greater part of Canada, however, are barely 60 years old.

Air transport has been the main force in promoting collection of climatic data. Economic interests, especially those of agriculture, have also spurred progress in forecasting. The key to the understanding of so much of the continent's weather lies in areas peripheral to it. No single area has been more important than the Arctic and the last generation has witnessed the rise of high-latitude observatories, such as those in Eureka Sound (Ellesmere Island) and Resolute Bay (Cornwallis Island). These may be either manned or automatic. North American interest in climatology has bred simultaneously a new generation of systematists. Among them, C. Warren Thornthwaite has constructed one of the most widely accepted systems of climatic regions.

OROGRAPHY AND MOVEMENTS OF AIR

Climatic variety is a function of the position and extent of the continent, its surface configuration and the waterbodies which bound it. With a latitudinal spread of 10 to 75 degrees north of the equator, it is inevitable that there should be a range of climates from the Tropical Savanna to the Tundra. Of the major climatic types, only the Tropical Rainforest is not represented. Because the base of the land triangle is in the north, cold and cool climates predominate. The warm southern apex suffers two disadvantages: much of it is desert or semi-desert and the Gulf of Mexico takes a large bite out of what might otherwise be a climatically favoured area.

The north-south orientated Cordillera, comprising the Coast Ranges and Sierra Nevada-Cascades, the Interior Plateaux and the Rocky Mountain system, is a formidable barrier to any west-east or east-west movements of air. The Pacific coast ranges take the full force of the impact of moist Pacific air and rob it of its moisture, leaving high orographic rainfall on the seaward slopes and rain shadow belts in their lee. These features are especially marked north of 40°N where onshore winds prevail throughout the year. Pacific influences are almost entirely excluded from the interior of the continent. The Rockies form a second line of defence, but so effective are the main defences that they are rarely called into action. Their principal climatic function is as a barrier to the westward movement of cold winter air from the interior. This they perform well, so that the Pacific coastlands very rarely feel the icy finger of continental conditions. North America, in contrast to Europe with its east-west grain, has no air funnels open to the west.

Again in contrast to Europe, North America has a great north-south funnel. From the Arctic Ocean to the Gulf of Mexico there is no latitudinal physical obstruction to longitudinal movements of air. The Mackenzie and Hudson Bay drainage basins are nowhere separated from their mirror image, the Missouri-Mississippi system, by ground high enough to obstruct the free passage of air in either direction. Along this draughty corridor Arctic influences push southwards, occasionally as far as the Gulf coast, whilst moist sub-tropical air from the Gulf may penetrate northwards, certainly into southern Canada if not into the Arctic. In winter, north of 45°N, the snow-covered surface of the corridor, dominated by a continental high-pressure cell, is the source region of cold, dry continental Polar air (cP). Body blows of maritime polar air from the Pacific which might sap its strength are excluded by the Cordillera.

The Appalachians are not a barrier obstructing air movement, rather they seem to channel the easterly flow of depressions north-eastwards up the Ohio and down the St. Lawrence valleys. There is no marked rain-shadow effect on the eastern Piedmont.

The Pacific Ocean north of 45°N is in winter the home of the Aleutian low-pressure cell—in reality an integration of a large number of migratory depressions. The prevailing wind circulation around this cell brings a south to south-westerly airstream to bear upon the coast of North America, north of Cape Mendocino, and with it the warm waters of the Alaska ocean current. In this Pacific mixing bowl cold Arctic air from the north and warm tropical air masses are so modified as to become a distinctive maritime Polar air mass. The character of this air when it impinges upon the coast depends upon the trajectory it has followed. If this has been from the north and the sea-passage short, it is wet, cold and unstable; if it has derived from tropical air and arrives on the coast after a long sea passage on a southerly trajectory it is also wet but cool and stable. In these two forms mP air usually makes up the cold and warm sectors respectively of depressions. By the time the depressions reach the coast they are deep and occluded. They give heavy and steady frontal and orographic rain on the coast mountains amounting to over 80 inches p.a. The combination of southerly onshore winds and the Alaska current keeps temperatures anomalously high for the latitude; at Vancouver the January mean monthly temperature is 36°F; at St. John's, Newfoundland, it is 24°F. But these characteristics are severely localized by the relief. Short distances from the coast in sheltered valleys rainfalls as low as 10 inches and January mean temperatures of 25°F are usual. Sometimes such depressions may struggle across the Cordillera and much modified mP air then descends on to the High Plains as the *chinook* wind. The temperature of this adiabatically warmed air far exceeds that of the indigenous cP air, as the figures opposite show:

Rapid City, S. Dakota 23.1.43

10.20 a.m.	5°F	12.30 p.m.	15°F
10.30 a.m.	42°F	12.45 p.m.	50°F
10.33 a.m.	20°F	1.00 p.m.	15°F
10.45 a.m.	50°F	1.30 p.m.	60°F
11.00 a.m.	50°F	5.20 p.m.	60°F
11.45 a.m.	60°F	5.30 p.m.	15°F
12 noon	13°F		

These figures reflect changes in the position of the boundary between the *chinook* and the cP air and also the passage of pockets of warm air enclosed within the cold air. Any snow cover is melted rapidly and the ground dries quickly under the influence of the warm blast.

OCEANS AND CLIMATIC DEPRESSIONS

South of 40°N the coast is washed by the relatively cold California current. North of San Francisco, in summer, cold bottom water wells up along the coast under the influence of northerly winds. Summer advection fogs are frequent where maritime air crosses the cold-water belt. They penetrate a short distance inland as fog or low-stratus cloud. Under these conditions daytime maximum temperatures are held down and this is reflected in the low summer mean monthly temperatures and annual range at coastal stations such as San Francisco. Depressions in winter have maritime tropical air in their warm sectors. Their occlusion is accelerated by the mountain ranges athwart their path if the process is not complete by the time they reach the coast. They are responsible for heavy frontal and orographic winter rain. George Stewart's *Storm* is a minor literary masterpiece of the human consequences of the passage of such a depression across California. In summer the Pacific sub-tropical high dominates the pressure pattern, consequently summer rainfall along the whole Pacific seaboard is low, nil in the Mediterranean climatic region south of 40°N.

Although the continent is open to attack from the east, the Atlantic has much less influence upon the climate of North America than the Pacific. The continental high-pressure cell of winter, around which anti-cyclonic circulation brings north-easterly winds to the eastern seaboard north of 35°N, causes continental conditions to prevail right to the coast.

The Icelandic Low has little influence. Along the same stretch of coast the Labrador cold current would accentuate the low winter temperatures were it not for the offshore winds which hold its effects at bay. The sea regularly freezes south to latitude 46°N in winter, an occurrence not known on the Pacific coast even as far north as 60°N. The Florida Current hugs the south-east coast as far north as Cape Hatteras, where the coastal trend changes from north-east to north and the current moves north-eastwards into the Atlantic as the Gulf Stream. The warm waters of this current have little influence upon the climate of the adjacent land mass. The Atlantic sub-tropical high-pressure cell minimizes air movement in winter, when the land is usually dominated by a continental air mass. Of much more consequence is the Atlantic's western antechamber, the Gulf of Mexico. The importance of this waterbody as a source of moisture cannot be exaggerated. If it were land a large part of the Mississippi Basin would be desert and semi-desert. It is the principal source of the maritime tropical air which invades the continent from the south. Warm moist air from the Gulf meets cold dry continental air along the prolongation of the Atlantic Polar Front and is uplifted in a continuous series of depressions which move eastwards across the land mass. Depending upon the relative strengths of the two air masses, the location of the Polar Front varies from a position along the Gulf and Atlantic coasts to one just inside Canada at 50°N. A common source of depressions is the Colorado piedmont. They follow a curved path, at first moving south-eastwards, gradually swinging north-eastwards to the west of and parallel to the Appalachians, and leave the continent via the St. Lawrence lowlands. Frontal rain from the uplifted mT air in winter, and convection rain in summer from the same air mass warmed from below by a hotter land surface, are the life blood of the habitable lands east of the Rockies. Anticyclones form where cP air in the rear of a depression pushes south. These also migrate eastwards as new depressions are formed to their west. But the most frequently occurring depressions originate in the Rocky foothills of Alberta which may be Pacific depressions reborn after the Cordilleran crossing. They move rapidly eastwards and give light or moderate precipitation all the year.

Hudson Bay has a considerable influence upon the climate of the continent. In summer it exerts a cooling effect and is responsible for the cloudy summers of much of northern Canada. In winter when most of it freezes over, it acts as a land surface and greatly augments the source region of cP air.

The Great Lakes, although heavily iced around their shores, are rarely completely frozen over. They warm air in passage over them in winter and cool it in summer and are a ready source of moisture for dry continental air masses. These influences are expressed in the frequent forecasts of snow flurries in winter for the southern shores of the Great Lakes, when cP air dominates the scene, and the protection from late spring frosts on the south shores of lakes Ontario, Erie and Michigan. Their cooling influence can be most marked. On 25th April 1954 downtown Chicago near the lake front had a temperature of 53°F; six miles away at the airport, inland, it was 76°F.

CLIMATIC HAZARDS

The climates of North America are nothing if not exciting. To obvious contrasts from place to place may be added climatic hazards (Figure 8). In the coastal states of the east and south the period from August to November is the hurricane season. These tropical revolving storms with winds of 75 knots and over, but calm central eyes, accompanied by torrential downpours, originate in the Caribbean Sea and move north and west towards the U.S.A. Their individual courses are unpredictable, and a close watch is kept upon them by patrolling aircraft. As they impinge upon the land, the wind and rain inflict great damage. They are accompanied by wind-whipped sea-waves which, along the coast, may wreak even more destruction. Owing to friction and dependence upon a source of moisture, they quickly die out over the land. A number turn north-east, follow the eastern seaboard and may keep several million people speculating on their next move for several days on end. The Weather Bureau obviously regards their capriciousness as a feminine characteristic, for it christens each hurricane with a girl's name, in alphabetical order by initial letter each year. No film-star's fortunes are followed with as much interest as those of Ada, Betty, Carol and Dora. Hurricanes share with film stars the characteristic that the worse their behaviour the greater the ephemeral interest in them. It is indicative of their number that the Gertrudes of this world are fairly safe from invidious comparison.

Tornadoes, a few hundred yards across, are much more localized centres of low pressure. They occur most frequently in spring and early summer in association with thunderstorms. The high wind speeds and explosive effect of intensely low pressures cause great damage as they race across the country at speeds of 30 m.p.h. or more. They are marked by black funnels descending from cumulo-nimbus clouds, and occur most often in the plains between the Rockies and the Mississippi. Dorothy's adventures in the *Wizard of Oz* all started with a Kansas tornado which whisked her away (together with her Uncle Henry's wooden house) to the Emerald City before she had time to dive into the cellar.

The south-eastern United States is the most thundery area in the continent, with thunderstorms 60 or more days per year. These occur

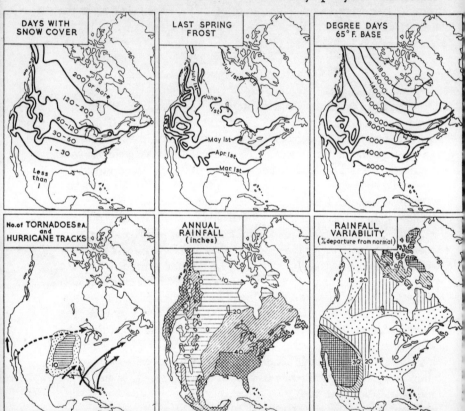

Figure 8 *Climatic Hazards*

A degree-day is a temperature of 1°F for 1 day above or below a given base temperature. Using a 65°F base the map is a measure of the extent to which temperatures fall below 65°F and therefore of the amount of heating required to maintain the temperature—65°F is a conservative estimate of room temperatures in North America. (Several sources including S. S. Visher, *Climatic Atlas of U.S.A.*, New York, 1958)

principally in maritime tropical air in summer, resulting from frontal uplift, local and advective surface heating. A high frequency in the southern Rockies (50–70 days p.a.) is related to orographic uplift. Hail is a product of such storms and insurance against its damage to crops is common.

The cold wave, meteorologically a southerly outburst of cP air in the rear of a depression, is a winter phenomenon. It is accompanied by a sharp drop in temperature resulting in içy roads, snowdrifts, cracked cylinder blocks and springtime destruction of fruit blossoms. A cold wave in the northern states is defined by the Weather Bureau as a drop of 20°F in 24 hours and a minimum of 0°F in winter; for Florida, a drop of 16°F and a minimum of 32°F. The *chinook* may have a dramatically contrasting effect. There are recorded instances of temperature increases of 15°F in five minutes.

TYPES OF CLIMATE

The ten climatic types defined on Figure 9 may be grouped into five fundamental divisions, of which two are conducive and three adverse to human habitation. The humid south-east, south of 50°N and east of 100°W, has a rainfall which is generally sufficient for crop growth. The length and severity of the winter and the related length of the growing season are the principal differences within it. Along the west coast is a narrow, humid strip with a summer deficient in moisture and a winter relatively inadequate in warmth. The areas of real climatic hardship are the cold half of the continent north of 50°N, and the arid south-western quadrant, subdivisions of which are based upon degrees of coldness and aridity respectively. Thus the following types of climate will be noted:

A. *Areas of Climatic Increment*

    Ai. The humid South-East
        (1) The tropical Tip
        (2) The humid sub-tropical South
        (3) The humid continental Interior

    Aii. The humid West Coast Margin
        (4) The North American 'Mediterranean'
        (5) The temperate oceanic Fringe

B. *Areas of Climatic Hardship*

    Bi.   The Cold Northlands
          (6) The Taiga
          (7) The Tundra or Cold Desert

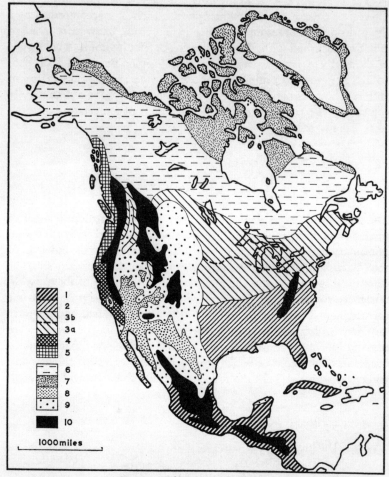

Figure 9 *The Pattern of Climates*

1 Tropical Tip. 2 Humid Sub-Tropical South. 3 Humid Continental
Interior. 4 North American 'Mediterranean'. 5 Temperate Oceanic Fringe.
6 Taiga. 7 Tundra. 8 Hot Desert. 9 Steppe. 10 Mountain. See text, pp. 75–
85. (Largely based on G. T. Trewartha—after Köppen—*An Introduction to
Weather and Climate*, New York, 1943)

Bii. The Dry South-West
    (8) The Desert
    (9) The Steppe
Biii. (10) The Mountains

Areas of Climatic Increment

Ai. The humid South-East (1) *The tropical Tip*
Florida, south of Daytona, experiences a true tropical climate as the following data show:

Miami O ft. A.S.L.

	J	F	M	A	M	J	J	A	S	O	N	D	
°F	67	68	71	74	78	80	82	82	81	78	73	68	15° range
"R	2·5	2·0	2·1	3·1	6·1	7·0	5·4	6·1	8·5	8·3	3·0	1·7	55" annual

There is no cold season, but in the coolest part of the year, November to April, rainfall is lower than in the hot season. December has less than 2 inches of rain, but the soil is never without moisture. Too much rainfall, most often associated with hurricanes in September and October, is the chief hazard. Frost is extremely rare, and the screen thermometer will, on average, only fall below 32°F on one night annually. The lower precipitation and high, but not excessive, temperatures prevailing from November to April offer a refuge from the cP air of the northern winter and are the physical base of the Florida tourist industry. Massive advertising campaigns and the offer of tantalizing hotel reductions to create a summer season have to combat not only high rainfall and temperatures but also high relative humidity. The struggle between climate and 'admass' is unresolved.

(2) *The humid sub-tropical South*
The same pattern of climate prevails in the south-eastern continental margins. Mean monthly temperatures are always above 32°F, but daily temperatures in winter may fall lower. There are over 200 frost-free days. In summer mean monthly temperatures exceed 70°F, but they are rarely extreme. Very low minimum temperatures may occur under cold-wave conditions when the cP air mass pushes south behind a migratory depression. At Montgomery (Ala.) the January mean daily minimum is 39.6°F but under cold-wave conditions the thermometer has dropped to –5°F.

    Rainfall is high, well distributed throughout the year and shows a summer maximum. Winter rain is principally cyclonic in origin, while the summer maximum reflects the convective uplift of mP air and the high frequency of summer thunderstorms. The number of days with snow on the ground ranges from nil in the south to ten in the north.

High relative and absolute humidities are characteristic throughout the year. These are bearable in winter; but in summer, when combined with the high temperatures, they are distinctly unpleasant. Air conditioning is then a boon, but the wet blanket of air which must be traversed between air-conditioned home and air-conditioned office (unless the car is also air-conditioned) is that much worse by comparison. Washington was at one time classed as a tropical assignment for British diplomats and a hard-living allowance paid.

The summer heat, if not the humidity, may be escaped in the high ground of the Appalachians where, for the same orographic reason, winter temperatures are also lower, snow cover longer and precipitation generally higher.

The principal climatic hazards are the cold wave, the thunderstorm, the tornado in the west, and the hurricane in the coastal areas. These aberrations, important as they are, are not revealed by climatic averages such as the following:

Charleston N.C. O ft. A.S.L.

	J	F	M	A	M	J	J	A	S	O	N	D	
°F	49	52	57	64	72	79	81	80	76	67	58	51	32° range
"R	3·1	3·1	3·3	2·4	3·4	5·3	6·2	6·7	5·2	3·9	2·7	3·3	48·6" annual

(3) *The humid continental Interior*

The humid continental interior type of climate can be subdivided according to whether it is characterized by warm or by cool summers (3a and 3b respectively on Figure 9.) Long, hot, sultry summers are features of the following climatic station:

Indianapolis 400 ft. A.S.L.

	J	F	M	A	M	J	J	A	S	O	N	D	
°F	28	30	40	52	63	72	76	74	67	55	41	33	48° range
"R	3·0	3·1	4·0	3·4	4·0	4·2	4·1	3·2	3·0	2·7	2·5	3·0	41·2" annual

Rainfall occurs throughout the year, but there is a clear summer maximum. The annual total is remarkably high for a continental climate without orographic help, but the procession of winter depressions brings cyclonic rain and the mT air from the Gulf of Mexico is a source of summer convectional rain. Actual winter temperatures fluctuate rapidly as the front between cP and mT air crosses the region. The number of days with a snow cover is a good index of the impact of winter. Chicago has 35 inches of snow p.a. with 50–60 snow days; St. Louis, 300 miles to the south, has 20 inches but only 14 snow days. St. Louis can live in spite of the snow, Chicago has to live with it and make due allowance for almost 2 months of it.

North of Chicago the climate changes imperceptibly into a humid continental climate with cool summers. The contrast with the region of warm summers is expressed in the following data:

Winnipeg (Man.) 786 ft. A.S.L.

	J	F	M	A	M	J	J	A	S	O	N	D	
°F	–4	0	15	38	52	62	66	64	54	41	21	0	70° range
"R	·9	·7	1·2	1·4	2·0	3·1	3·1	2·2	2·2	2·2	1·1	·9	20·2" annual

Although summer temperatures are lower by some 10 degrees, there is a greatly increased annual range. This reflects much lower winter temperatures, in which the difference is about 30 degrees—from just below freezing to just below zero. The growing season is some 3–5 months in contrast to the 5–7 months in the region south of the Great Lakes. Extremes of –30°F to –50°F are experienced and survived quite happily with the help of dry crisp air and winter sun. The low precipitation, comparable to that of London, is least in winter when it falls as snow and covers the ground from 3 to 5 months. In summer, when the maximum is experienced, it falls as convectional rain with occasional thunderstorms accompanied by hail of golf-ball dimensions. In this 'Middle Western' area, spring in the West European sense of the term does not appear in the climate. Winter temperatures below freezing alternate with mid-summer temperatures in the 70's, trees may burst into full leaf in several short weeks, and be shrivelled by frost. It has been facetiously said that the Middle West lies 'outside the latitudes of love'.

East of the Great Lakes the extremes are not as great, as the data overpage for Montreal shows:

Montreal (P.Q.) 187 ft. A.S.L.

	J	F	M	A	M	J	J	A	S	O	N	D	
°F	13	15	25	41	55	65	69	67	59	47	33	19	56° range
"R	3·7	3·2	3·7	2·4	3·1	3·5	3·8	3·4	3·5	3·3	3·4	3·7	40°7" annual

The annual precipitation is much higher, largely because winter precipitation, much of it snow, is far higher. It derives from a much larger number of depressions which, wherever their Rocky piedmont origin may have been, seem to depart east by way of the St. Lawrence estuary. Winter temperatures are consequently higher by 10 to 15 degrees, but the atmosphere is noticeably raw. The orographic influence, where present, serves to increase the amount of winter snow to the advantage of the skier, and to decrease summer temperatures to the relief of the vacationing southerner.

Aii. The humid West Coast Margin (4) The North American 'Mediterranean'

Sacramento 71 ft. A.S.L.

	J	F	M	A	M	J	J	A	S	O	N	D	
°F	46	50	54	58	63	69	73	72	69	62	53	46	27° range
"R	3·8	2·8	2·8	1·5	·7	·1	0	0	·3	·8	1·9	3·8	18·5" annual

These figures are representative of inland sites in California which are not unduly influenced by the effect of altitude or topography. The temperatures are sub-tropical and comparable to those of the south, but the rainfall régime is quite different. It is a 'Mediterranean' type climate in which rainfall is largely confined to the period October to May. In summer there are up to 4 months of absolute drought and the climate resembles that of the adjacent desert.

The influence of altitude in this region is shown in the climatic data on Figure 46. The effect of shelter from oceanic influences is well illustrated by Red Bluff, at the northern end of the Central Valley of California east of the Coast Ranges, and Sacramento, also in the Central Valley but open to the Golden Gate. Red Bluff experiences a 45°F mean January temperature, an 82°F mean July temperature; Sacramento, a 49°F mean January and a 73°F mean July. The difference of 1½ degrees in latitude plays some part, but it is subordinate to the continental effect. Frost is a climatic hazard. Red Bluff with 10 to 12 frosty nights p.a. has approximately twice as many as Sacramento; Los Angeles, farther south, experiences a frost approximately 1 year in 2. Tinder-dry conditions in the latter half of the summer give rise to the seasonal hazard of brush and forest fires.

The major variation within the Mediterranean type of climate results from proximity to the coast:

San Francisco 207 ft. A.S.L.

	J	F	M	A	M	J	J	A	S	O	N	D	
°F	49	51	53	54	56	57	57	58	60	59	56	51	11° range
"R	4·8	3·6	3·1	1·6	·7	·1	0	0	·3	·9	2·4	4·5	22·7" annual

The rainfall régime is the same as Sacramento, but the annual total is somewhat higher, as befits a coastal site. Winter temperature differences can be similarly explained, but there are startling differences in summer temperatures. The lowering of the summer maximum by 10 to 15 degrees and its delay until September is a consequence of sea fog which may shroud the coastal margin all day. Moist, warm air moves landwards across the cold-water zone close to the coast; it is cooled below its dew point, condensation occurs, advection fog rolls up the valleys and through the gap of the Golden Gate, before it is dispersed by heating from below by the warmer land surface.

Los Angeles suffers from its own variety of man-made fog or smog. On 340 days of the year a temperature inversion exists over the city; on 150 of them it is low enough (1,500 feet) to trap the air pollution generated by a city of several million cars. This gives rise to a brownish-yellow haze which is painful to both eyes and nose.

(5) *The temperate oceanic Fringe*
This type of climate is the North American equivalent of that in Britain:

Victoria B.C. 85 ft. A.S.L.

	J	F	M	A	M	J	J	A	S	O	N	D	
°F	39	40	43	48	53	57	60	60	56	50	45	41	21° range
"R	4·5	3·2	3·5	1·6	1·2	·9	·4	·6	1·8	2·5	5·7	5·8	30·9" annual

F

The temperature régime is completely representative of its type and closely resembles that of Dublin. Rainfall occurs in all months, but its distribution is atypical in that June, July and August each have less than 1 inch of rainfall. This statistical guarantee of a good summer helps to make the climate 'feel' better than its western-European counterpart. Nowhere in the British Isles has any month an average of less than 1 inch of rain. April is usually the lowest, when London, for instance, has 1·5 inches.

Dry summers are a Mediterranean characteristic; because the wettest winter month has more than three times the rainfall of the driest month the climate is classified as 'Mediterranean' by Köppen. The annual rainfall is too high for this to be a happy designation, so that the limit between the Mediterranean and west-coast temperate marine climates is taken to be Cape Mendocino, slightly more than 40°N. The longitudinal relief induces a rapid rise in rainfall within short distances of the sea (Figure 46) and limits the climate strictly to the coastal fringe. Local climatic influences, particularly aspect to onshore winds, play an important role. Summers are certainly cooler on the Alaskan coast. Anchorage has a July mean of 57°F; but, because of the influence of the Alaska current, this is much higher than might be expected for a station 10 degrees of latitude north of Victoria. Winter rainfall from Pacific occlusions is high.

Areas of Climatic Hardship

There are two areas of climatic hardship—one in the north, with its affliction of cold; the other in the south and west, with its problem of drought.

Bi. The Cold Northlands (6) The Taiga

The most extensive climate in the whole continent is the one with the greatest temperature extremes:

Dawson (Yukon) 1,062 ft. A.S.L.

	J	F	M	A	M	J	J	A	S	O	N	D	
°F	-23	-11	4	29	46	57	59	54	52	25	1	-13	82° range
"R	·8	·8	·5	·7	·9	1·3	1·6	1·6	1·7	1·3	1·3	1·1	13·6" annual

In this land of coniferous forest summers are short but the days may be long and temperatures may rise into the 80's. Outdoor life is attractive, but not revealed by climatic statistics is the menace of the mosquito against which the use of smoke, oils or wire screens is imperative. A growing season of some 50 to 75 days, as in the MacKenzie Valley, is possible; but the long winter is very cold and a blanket of snow 2–3-foot deep covers the earth for 5 to 7 months.

(7) *The Tundra or Cold Desert*

Point Barrow (Alaska) 22 ft. A.S.L.

	J	F	M	A	M	J	J	A	S	O	N	D	
"F	–19	–13	–14	–2	21	35	40	39	31	16	9	–15	59°
													range
"R	·3	·2	·2	·3	·3	1·1	·8	·8	·5	8	·4	·4	5·6″
													annual

The brief summer, with mean monthly temperatures above 32°F, is also the period of maximum precipitation in the cold desert lands. The precipitation is therefore mostly rain, but snow is not uncommon even in summer. Everything combines to make summer cheerless and raw; rain, low temperatures, wet earth, grey cloudy skies, fog near open water surfaces, and mosquitoes. The sun, when visible, is above the horizon for all or most of the 24 hours. Winters are bitterly cold, snowfall is light; but winds are strong and blizzards sweep across the open plains.

Bii. *The Dry South-West* (8) *The Desert*

The hot and cold deserts have in common very low annual precipitations, but there the resemblance ceases.

Yuma (Arizona) 140 ft. A.S.L.

	J	F	M	A	M	J	J	A	S	O	N	D	
°F	55	59	65	70	77	85	91	90	84	72	61	56	36°
													range
"R	·5	·4	·3	·1	0	0	·2	·6	·3	·2	·3	·4	3·3″
													annual

Hot and dry summer days with temperatures rising to between 105 and 115 degrees in the afternoons are a severe strain upon the human body. A drop of 20 to 30 degrees at night is a blessed relief and seems cold by comparison, but no one can pretend that 80 degrees is cold. Record temperatures reach over 130°F. The heat is accompanied by almost relentless sunshine, 80 to 85 per cent of that possible per year, only 20–50 days in a year with cloud, 12 to 13 hours sunshine per day in summer, 7 to 8 in winter. Rainfall hardly matters, an occasional thunderstorm in summer, a very little cyclonic rain in winter. Variability is its most conspicuous characteristic. Averages are meaningless.

(9) *The Steppe* (10) *The Mountains*

Transitional to the hot deserts on the one hand and to the humid east on the other are the semi-desert steppes of the intermontane plateaux and the High Plains. Within the steppe the Rockies have their own altitudinally generated mountain climate. Because of its location the steppe climate is also markedly influenced by altitude. No fixed boundary exists; there is a gradual transition between the desert and the semi-desert. This type of climate extends from Alberta to Mexico and it is necessary to distinguish subdivisions. A division into sub-tropical and mid-latitude steppe is also made after Köppen, the distinction being whether the temperature of the coldest month is above or below 32°F.

The sub-tropical steppe is characterized by data for Albuquerque in New Mexico near the southern limit of the Rockies:

Albuquerque (N. Mexico) 5,200 ft. A.S.L.

	J	F	M	A	M	J	J	A	S	O	N	D	
°F	35	40	48	55	64	73	76	74	68	56	44	35	41° range
"R	·4	·2	·2	·5	·4	·8	1·2	1·3	·9	·7	·5	·4	7·5" annual

The rainfall régime is continental, with a summer convectional maximum. Temperatures are depressed all the year round by the altitude so that summers are pleasantly cool and winters rather cold for the latitude.

The nature of the mid-latitude steppe climate in a comparable situation is shown by the data for Helena, Montana:

Helena (Montana) 4,100 ft. A.S.L.

	J	F	M	A	M	J	J	A	S	O	N	D	
°F	20	23	32	44	52	60	68	67	56	45	33	25	48° range
"R	·9	·6	·8	1·1	2·1	2·3	1·1	·7	1·2	·9	·7	·8	13·4" annual

Snow is clearly significant in the winter climate of this region.

The climate of the High Plains must be examined more closely because it is transitional to the humid climates. On the High Plains the *chinook*, the cold wave and the depression are characteristic, but the most important single feature is rainfall variability. One year in three rainfall is less than 85 per cent of normal. Minot (N. Dakota) has a mean annual precipitation of 15·2 inches but between 1892 and 1930 the highest was 24·8 inches, the lowest 7·2 inches. In some years, therefore, rainfall is sufficient, if not abundant, for crop growth: in others, it is grossly deficient.

In other words, the region of semi-arid climate east of the Rockies has a humid climate in some years, whilst in others it verges on the desert. On the basis of the mean for 35 years the boundary between the humid and semi-arid lands is located approximately along the 100° meridian; but in wet years it is coincident with the Rockies 200–300 miles farther west. There is therefore a meridional belt some 600 miles wide within which rainfall is extremely variable and farming a gamble. The 100° meridian is an important line in the geography of North America as it also marks approximately the western limit of chernozem soils and the transition from long to short-grass prairie (Figure 41).

A SURVEY OF SOILS

Contemporary soil studies in North America spring essentially from the work of C. F. Marbut, who as a young man from the Missouri Ozarks drew his inspiration jointly from the Russian School of Dokuchaiev and Glinka and his American forerunner, Hilgard. The generally accepted zonal soil groups of North America are identified on Figure 10 and in the ensuing text the bracketed numerals refer to the groups marked upon it.

The soils of the cold zone embrace those of the *Tundra group* (i). They commonly assume the form of a peaty surface layer covering a blue-grey sticky and compact subsoil. Since bedrock and, sometimes, subsoil are permanently frozen, soil drainage is hindered and boggy conditions prevail when soils thaw in summer. Muskeg peat soils are found in depressions.

Light-coloured podsolized forest soils usually succeed the soils of the cold zone on their favoured margins. They may also be termed pedalfers, a word composed of the initial letters of pedology, alumina and ferrous, indicative of their principal mineral constituents. They are thus contrasted with the pedocals, soils in which calcium carbonate is an important mineral element. Carbonates have been removed from the pedalfers by strong leaching action. Several subdivisions of the podsolized soils may be identified. True *podsols* (ii) are commonly developed under a cover of coniferous woodland. There is an abundant surface litter of acid mor,

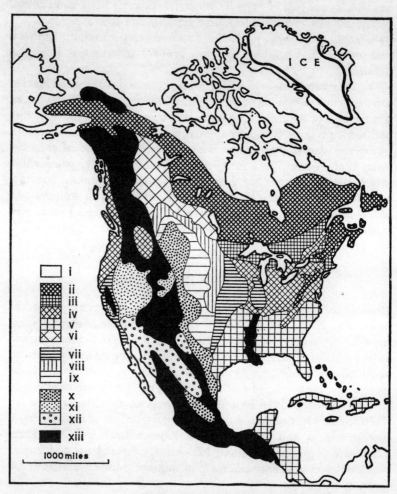

Figure 10 *Types of Soil*

i Tundra. ii Podsol. iii Brown Podsolic. iv Grey-Brown Podsolic. v Red-Yellow Podsolic. vi Grey Wooded Podsolic. vii Prairie. viii Chernozem. ix Chestnut. x Brown. xi Sierozem. xii Red Desert. xiii Azonal. The figures are referred to in the text, pp. 85–90

derived from the slow decay of the needles. Earthworms and bacteria are of little importance in this acid environment. Leaching is high, productivity low and once the organic matter has been exhausted fertility slumps, as the ash-grey A_2 horizon beneath is devoid of plant nutrients.

Southwards the process of podsolization becomes less extreme so that the podsol soils gradually merge with the *brown podsolic* (iii) soils of the southern fringe of the Canadian Shield and of New England. They are acid because they were formed under mixed coniferous and deciduous forest; but they are better suited for agriculture than the podsols. They soon yield to *grey-brown podsolic soils* (iv) developed under deciduous trees. A mull layer several inches thick overlies a grey or yellowish-brown platy horizon which still contains some calcium so that the soil is not excessively acid. These soils are of medium quality; but because of the particular conditions of rainfall, temperature and length of growing season under which they occur they have become vital to American agriculture (their tree cover cleared). They extend in a great arc from the St. Lawrence lowlands west to Iowa and south beyond the limits of glaciation. As with all soil groups the grey-brown podsolic is made up of several soil *series* each derived from the same parent material. Some of the principal series developed upon the glacial drift of the Middle West are the Miami series on Wisconsin till, the Belfontaine series on Wisconsin kame morainic deposits, the Cincinnati series on Illinoian till and the Princeton series on Wisconsin loess. Even a single soil series includes a wide range of soils and it is usual to divide each series into soil *types* on the basis of the texture of its A horizon, e.g. the highly fertile Miami silt loam, a soil occurring on the well-drained but gentle slopes of undulating uplands on calcareous Wisconsin (Tazewell) glacial till. Even the soil type can be subdivided into *phases* based upon departures from the norm of the type. Thus the Miami silt loam, sloping phase, occurs on medium slopes, is well to extremely well drained, is thin and of medium fertility. Large areas of the grey-brown podsolic soils also occur in the coastal and valley regions of northern California, Washington and Oregon.

In the southern states, under a humid sub-tropical climate and a mixed forest cover, *red-yellow podsolic soils* (v) are found. Their colour is due to the hydrated state of the iron salts in the mineral matter. The virgin soil has a thin layer of unincorporated organic matter overlying, in turn, 2 to 3 inches of humus incorporated soil in the A_1 horizon, a yellowish-grey A_2 horizon of several inches thickness, and a red B horizon. Farther south under still wetter and warmer conditions the B horizon is yellow. The total depth of weathering reaches several feet.

Injudicious exhaustion of their fertility has led to the removal of the humic A horizon by sheet and gully erosion. Cultivation of the remaining B horizon results in an 85 per cent loss of productivity, while transported soils choke reservoirs and aggravate floods.

There are, finally, the *grey wooded podsol group* (vi), which occur principally between the Canadian Shield and the Rockies. They extend in an arc outside the grassland soils of the Prairies as a belt of soils formed under the discontinuous tree cover of the Aspen Parkland (see Figure 11). Climatically there is a moisture deficiency, so that although leaching is evident the soils are not very acid and there is a strong tendency for calcium carbonate to accumulate in the C horizon. They are considerably lower in fertility than the adjacent chernozems but their low acidity contrasts markedly with their other neighbours, the strongly acid podsols, which they otherwise resemble.

Dark-coloured grassland soils form the third main zonal soil group and occur extensively. They include the prairie, chernozem and chestnut soils.

The *prairie soils* (vii) are also podsolic, as the climatic conditions under which they were formed closely resemble those prevailing in the region of the grey-brown podsolic group. Prairie soils were developed under long grass, and are high in fertility. The dark-brown granular A horizon is high in humus and, although acid in reaction, the leaching is not high. Southwards, the soils pass into reddish prairie soils (not separately distinguished on Figure 10). There is lively debate as to whether the grassland soils ever supported woodlands, or could again if undisturbed by man.

The black soils of the *chernozem group* (viii) have been formed under mixed grassland with a rainfall of 15 to 20 inches annually. Calcium carbonate leached out of the top layers accumulates in the upper C horizon. Accumulation of decaying grass, low rainfall and hot summers have produced an A_1 horizon over a foot thick, black in colour, granular in structure and neutral to slightly acid. The chernozems are inherently the most fertile soils in North America, but because of the uncertain rainfall they are not as reliable in yield as the prairie soils.

West of the belt of chernozem soils, as rainfall decreases, leaching also decreases and the level of calcium carbonate deposition rises in the soil. At the same time grasses become shorter and the humus content of the soil diminishes. This is the zone of *chestnut soils* (ix). Traced south they become reddish in colour.

The arid regions support light-coloured soils which form a fourth zonal group.

Developed under more arid conditions than the chestnut soils and below a short, bunch-grass vegetation cover are the *brown soils* (x). On the High Plains they represent the end member in a gradual transition from prairie through chernozem and chestnut soils. Gypsum present in the subsoil was worked as a fertilizer but it has been replaced by superphosphate. Both the chestnut and the brown soils are extremely susceptible to wind erosion when the surface is broken and the shallow humic horizon depleted of its binding humus.

Sierozems (xi), which are grey in colour and developed under a discontinuous cover of sagebrush, characterize the intermontane and semisteppe. The *red desert soils* (xii) are developed under tropical-desert conditions with an even less continuous vegetation cover of creosote bushes and allied drought-resistant plants. A layer of calcium-carbonate deposition immediately below the surface horizon is common to both. In the red desert soils it is frequently cemented into a hard pan. Such soils are of little agricultural value until irrigated.

Intrazonal soils of varying character may also be identified. Such soils are usually dominated by local features and never develop the group character controlled by climate. In North America this arises principally through excess of moisture in the soil—either as a result of impeded internal soil drainage or a low-lying situation. In the podsolized soil group of the forested eastern lands, such soils are characterized by thick developments of peat which, when decomposed, form the graphically named and highly fertile muck soils. It is not only in industrial England that muck means money. Mineral soils, equally poorly drained, have glei horizons.

The prairie-soil areas also contain upland wet prairies, which result from lack of surface slope on the particularly smooth till plains. Wet prairie soils may be of either the wiesenboden or the planosol type. The former are due simply to poor internal drainage and when drained artificially are as fertile as the zonal prairie soil they otherwise resemble. The wetness in the planosol condition arises from the existence in the B horizon of a heavy clay pan, a product of podsolization. They are common in southern Illinois, Iowa and northern Missouri.

Soils with an abnormally high lime content, called calcimorphic soils, are well developed locally. The rendzina, a thin, dark grassland soil developed from limestone, occurs in the blacklands of northern Texas and in the black belt of Alabama. Similar brown forest soils developed under forest occur where the Niagara limestone outcrops between Lakes Winnipeg and Manitoba and on highly calcareous till in southern Ontario.

Azonal soils (xiii) comprise thin soils on steep slopes (as in the western Cordillera), immature soils on recent deposits of sand, loess and glacial drift, and alluvial soils along such valleys as the Mississippi and in the Central Valley of California.

THE NATURE OF THE VEGETATION

In few countries is the public conscience made so continuously aware of its biological resources as in the U.S.A. and Canada. The principal holders of this conscience have been the Federal Forest Service in the U.S.A. since 1879, and the Forestry Branch in Canada since the 1890's, together with State and Provincial forestry interests. Conservation of what remains and restoration of that destroyed have succeeded to the age of exploitation.

In spite of three and a half centuries of clearance trees still dominate the North American scene. Only in the tundra, the great plains (save for river valleys) and the western intermontane scrub lands (save for mountain ranges) are trees not naturally dominant.

Figure 11 offers a subdivision of woodlands according to the predominant species of trees. The close relationship between climate, soil and vegetation may be seen by comparing Figures 8, 10 and 11.

The boreal forest or *Taiga* (1) belongs to the circumpolar belt of coniferous forest and its limits coincide approximately with those of the sub-arctic climate and podsol soils. Its principal constituents are black and white spruces and balsam fir, the main sources of pulp; jack pine and tamerack (larch), which creosote well, are resistant to decay and are used for poles, posts and railway sleepers. The trees vary in height according to age and soil conditions, but average 30–50 feet. The forest is not a high one. Pure stands of a single species are common and tree crowns are contiguous.

Northwards the closed forest yields to an open, sub-arctic *Parkland* (2), the 'land of the little sticks', where trees are smaller as well as more widely scattered. Although peat bogs are common the parkland is better drained than the boreal forest. The final stage in the transition from forest to tundra is through what the Russians call the 'forest tundra' ecotone, a landscape of forested valleys with stunted birch and scrub alder fingering between broad expanses of true tundra.

In the Prairie Provinces there is another type of transition from boreal forest to grassland, through a belt of *Aspen Parkland* (3). This open woodland, associated with soils of the grey-wooded group, is dominated by the deciduous trembling aspen and spruce.

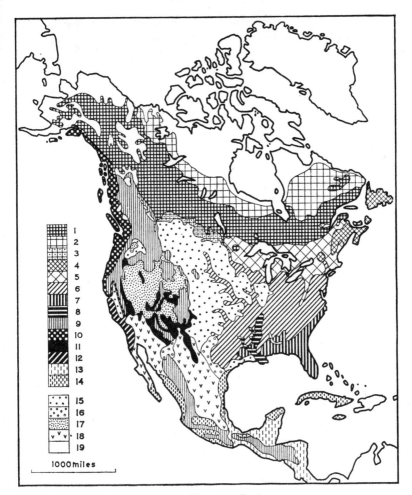

Figure 11 *Vegetation Regions*

1 Boreal Forest. 2 Sub-arctic Parkland. 3 Aspen Parkland. 4 Acadian Forest.
5 Great Lakes-St. Lawrence Mixed Forest. 6 Deciduous Forest. 7 South-
Eastern Pine Forest. 8 Bottom and Forest. 9 Western Pine Forest.
10 Pacific Coast Forest. 11 Pinon-Juniper Forest. 12 Chaparral. 13 Sub-
tropical Forest. 14 Sub-tropical Scrub Forest. 15 Temperate grass.
16 Mesquite. 17 Sagebrush. 18 Desert Scrub. 19 Tundra. The figures are
referred to in the text, pp. 90–5

Canadian foresters recognize the *Acadian Forest* (4) as a separate type. It has affinities with the boreal forest in that white spruce and balsam fir are present. Hemlock, white and red pine, the conifers of the Great Lakes-St. Lawrence Mixed Forest are also present, as are some of the hardwoods of the deciduous forest, sugar maple, yellow birch and beech. It may also contain the red spruce. The podsol soils are deeper than those of the boreal forest and have been developed under a higher rainfall.

The trees of the *Great Lakes–St. Lawrence Mixed Forest* (5) are now largely cleared. The native forest consisted of white pine (here reaching its greatest development), hemlock, red pine, jack pine, sugar maple, yellow birch, beech and aspen. The associated soils are of the brown podsolic group. The forest extends on to the southern margin of the Shield, but it is confined to the area south of the 0°F January and 60°F July isotherms.

The range of climatic and soil conditions within the region of the *Deciduous Forest* (6) is great. Three principal types of forest are normally recognized, roughly aligned in belts which trend from north-east to south-west. The inner belt adjacent to the grasslands is oak-hickory forest. From its base in Ohio, Indiana, Missouri and Oklahoma it extends in ribbons along the valleys of the Mississippi and Missouri and their tributaries, through the grasslands almost to the forests of the Rockies. Conversely, islands of grassland, particularly the prairie island of Illinois, exist within the confines of the hardwood forest. The second belt comprises chestnut, chestnut-oak and yellow poplar forest. It extends from southern New England across the Appalachians as far as the Mississippi below the Ohio confluence. The southernmost belt is transitional in that it is a mixed forest where the co-dominant trees are oak and pine. It occurs from Texas to the Atlantic coastal plain in Virginia on red podsolic soils. On the coastal plain it is replaced by the pine-dominant forest of the pine barrens.

The principal species of the *South-Eastern Pine Forest* (7) are longleaf, loblolly and slash pines. The slash pine is a good pulpwood tree: the pitch pine of the coastal fringes remains a source of resin and turpentine. The belt is developed on the strongly leached yellow podsolic soils.

The wet valley floors in the coastal plain from Virginia to Texas are clothed in *Bottomland Forest* (8) with cypress, tupelo and sweetgum as the distinguishing trees.

The *Western Pine Forest* (9) extends over the main mountain ranges of the Western Cordillera, with the exception of the Coast Ranges and Coastal Mountains. Within the wide range of latitude involved there are many climatic variations, but three species are dominant—the ponderosa pine, the Douglas fir and the lodgepole pine. The first, which reaches heights of 160 to 170 feet, extends from southern British Columbia southwards. Its timber is of fine quality, is easily worked, and is used for house construction, crates, furniture and toys. The Douglas fir is the most important timber tree in the U.S.A. and Canada. The blue or Rocky Mountain variety attains heights of around 100 feet, and is used for poles, railway sleepers and general construction purposes. The slender lodgepole pine, with its small tufty crown, is a distinctive tree valuable as a pulpwood. In the Sierra Nevada and Cascades the sugar pine—largest of the pines— is co-dominant with the ponderosa pine; while in British Columbia the blue-sheened Engelmann spruce is a major constituent.

Altitude principally controls vegetation type in the Western Cordillera, and it operates through temperature, rainfall and aspect. There is consequently a clear altitudinal zonation of vegetation from Alpine tundra on the highest peaks through forest, grassland and sagebrush to desert scrub and perhaps salt scrub.

The characteristic tree at the southern limit of the *Pacific Coast Forest* (10) is the coast redwood, *Sequoia sempervirens*, the world's tallest tree, not to be confused with the *Sequoia gigantea*, the carefully preserved redwood of the Sierra Nevada, groves of which include the world's largest (in girth) and oldest trees. The most important tree in Washington, Oregon and Southern British Columbia is the giant Pacific Douglas fir. It is the largest tree in Canada, commonly attaining heights of 150 to 200 feet and diameters of 3 to 6 feet. On good sites it reaches over 300 feet in height and 15 feet in diameter. Its strong, straight-grained, yellowish wood is most important for heavy construction work, veneer and plywood, poles and piling. Other trees in the forest are hemlock, grand fir and western white pine, the leading source of matchwood.

North of Vancouver Island the dominant tree is the Sitka spruce, the largest of all spruces. This is seldom found more than 50 miles from salt water or higher than 1,000 feet. It is tall and straight and produces a light, soft, straight-grained wood suited well (in the past) for aircraft construction but also valuable as a pulp tree. The large slow-growing western hemlock is much used for general construction and also for pulpwood. It rarely occurs in pure stands, but is mixed with Douglas fir, Sitka spruce and the western red cedar.

The short scrubby open growth of conifers classed as *Pinon-Juniper Forest* (11) occurs between 5,000 and 7,000 feet, below the ponderosa pine zone of the western pine forest. Its lower limit of growth closely follows the 12-inch annual isohyet. The saw-timber forest above it indicates a mean annual total of 19 to 25 inches of rain and snow. This open, low forest between the colder, densely forested slopes above and the hotter, lower, treeless elevations has been the favoured settlement zone for Indians and Europeans alike. The pinon nut, now handled commercially on a small scale, was an important source of food for the Indian.

Chaparral (12), broad-leafed shrubby growth, often called brush, occupies some 11,000,000 acres in California where the annual precipitation is between 10 and 35 inches and temperatures do not exceed 100°F for long periods. It varies in character from a dense low stand of evergreen shrubs to a scattered oak-grassland, and includes large areas of cut-over timber that has reverted to brush. It is densest on north- and east-facing slopes. Greater insolation creates a drier environment on slopes with a south and westerly aspect and there the chaparral is thinner.

Sub-tropical Forest (13) and *Sub-tropical Scrub Forest* (14) are confined essentially to Mexico and Central America.

In the Middle West oak-hickory forest reluctantly yields pride of place to the tall grasses of the Prairie, one of the divisions of the *Temperate Grassland* (15). Periodic fires and, in places, a high watertable have inhibited tree growth where climate and soils would seem to be favourable. Elsewhere in the world, most prairie-type soils are tree-covered. The luxuriant grassland which the pioneers encountered was made up of tall coarse grasses such as blue joint and Indian grass with a strong flavouring of herbs. In spring it was a mass of flowers, in the fall it died back in a robe of rich red-brown.

A change from long to short grass occurs within the chernozem soil region. It is marked by a definite lightening of soil colour. During wet years long grasses are also in evidence west of the long grass/short grass boundary. They also push west on lighter soils whilst the short grasses push east on heavier ones. For these reasons a transitional belt of mixed prairie is often recognized by ecologists between the prairie proper and the short or plains grassland proper which extends from Alberta to Texas. Blue gramma, buffalo grass and June grass are the principal components of the latter.

Mesquite (16) is the loose name given to a form of desert-grass savanna, a scattered growth of thornbushes and cacti over a desert-grass cover.

Sagebrush (17) is a form of desert scrub dominated by the deciduous, silvery-green-leaved sagebrush. The plants stand about 3 feet apart and reach 2–6 feet in height. The density of the scrub varies with the maturity of the sierozem soils, which are often developed on extensive alluvial fans. Sagebrush indicates a pervious soil which is moistened to a depth of several feet and is free from alkali salts. It is good agricultural land under irrigation, can be dry-farmed in better-than-average years and grazed in all years.

On soils which are pervious and not unduly salty the dominant plant of the *Desert Scrub* (18) is the shiny, yellowish-green-leaved creosote bush. On heavier and saltier soils the desert saltbush is dominant. In the true salt desert, greasewood is the characteristic plant—except on the salt flats where no vegetation exists.

Amongst the striking vegetation forms in the desert are the giant cactus and the more common Joshua or Yucca tree. Except for the salt flats, the desert is not bare, but carries a considerable plant cover. When seen from the ground this cover may appear quite dense, but seen from above the wide spacing of the plants and the bareness of ground between them is evident.

A large number of plants grow in the thin soil which rests on the frozen subsoil of the *Tundra* (19) or cold desert. In the brief summer, wild bees able to pollinate the flowers are numerous; mosquitoes swarm in their millions over mosses, lichens, heathers, small berry-bearing plants and stunted willows, birch and alders and especially over swampy places where sphagnum grows. The vegetation is adapted to conditions of physiological drought.

The natural vegetational distribution has been profoundly modified by man. In the process, new vegetational distributions have resulted which have called forth new methods of interpretation. A classical expression of the agricultural regions that have largely succeeded to the natural vegetational regions was constructed by O. E. Baker, whose map is reproduced as Figure 12 (p. 97).

SELECTED READING

The geological basis of the continent is covered in
  P. B. King, *Evolution of North America* (Harvard, 1959)

The Geological Map published by the Geological Society of America is as invaluable as their Glacial Map.

The geology of Canada is summarized in
  Economic Geology Series, No. 1, Canadian Geological Survey (Ottawa, 1947)

The survey of Canada is given in
  Men and Meridians: a History of Surveying and Mapping in Canada, Vol. 2, 1867–1917, (Ottawa, 1967)

The standard works in physiography are

> N. M. Fenneman, *Physiography of the United States* (2 vols. New York, 1931 and 1938)
>
> W. W. Attwood, *Physiographic Provinces of North America* (New York, 1940)
>
> W. D. Thornbury, *Regional Geomorphology of the United States* (New York, 1965)

Both Washington's Weather Bureau and Ottawa's Meteorological Office publish daily weather charts.

A succinct account of the dynamic climatology of North America appears in

> F. K. Hare, *The Restless Atmosphere*—Chapter II (London, 1953)

Regional climatology is well covered. A valuable source is

> 'Climate and Man', *U.S. Year Book of Agriculture* (Washington 1941,)

A synoptic view of the climatological scene is given in

> C. F. Brooke, 'The Climatic Record', *A.A.A.G.* (1948)

A comprehensive and imaginatively provoking book is

> S. S. Visher, *Climatic Atlas of the U.S.* (Cambridge, 1954)

There is a steady stream, which threatens to reach flood dimensions, of publications dealing with the soils of the United States and Canada. A useful summary for the U.S. is

> *Year Book of Agriculture, Soils* (Washington, 1938)

A new soil classification has now come into use. It is outlined in

> *Soil Classification, A Comprehensive System, Seventh Approximation*, United States Department of Agriculture (Washington, 1960)

The pedogeography of Canada is summarized by

> D. F. Putnam, *The Geographical Bulletin*, 1 (1951)

The principal sources of information on vegetation are

> W. E. D. Halliday, *Forest Classification for Canada, Bulletin 89* (Ottawa, 1937)
>
> *U.S. Year Book of Agriculture, Trees* (1949) and *Grass* (1948)
>
> A. W. Kuchler, 'A Comprehensive Method of Mapping Vegetation', *A.A.A.G.* 45 (1955)

An example of a specialist study is

> J. D. Curtis, *The Vegetation of Wisconsin* (Madison, 1959)

The problem of the grassland is re-examined in

> R. T. Coupland, 'A Reconsideration of Grassland Classification in the northern Great Plains of North America', *Journal of Ecology*, 49, 1 (1961), 135–67

The origins of the grid-iron pattern in the American landscape are discussed in

> W. D. Pattison, *American Rectangular Survey System, 1784–1800* (Chicago, 1957)

Figure 12 *Agricultural Regions*
(after O. E. Baker, 'Agricultural Regions of North America', *E.G.*, 2, 459–93,
1926)

3

The Peopling of a Continent

A People without a Record—The Realm of the Amerind—The Landfalls of the Western World—The Political Definition of North America—The Frontier in Retreat—The Rise of an Urban Community—The Emergence of Canada—The Record of the People.

By European or Asian standards the occupation of the New World by man is very recent. It is the purpose of this chapter to consider its character for North America. In the process of occupation, Atlantic man has given the New World a new look. Yet, from time to time, the traveller through it is confronted with startling antiquity. It may be in the hawk eye or raven hair of the Indian, who is now largely found in reservations or absorbed into the hybrid American community. It may be in the 'mound buildings' of the Middle West (Ohio state alone has 6,000 of these earthworks), in the pre-conquest brick and adobe ruins of the Pueblo country of Arizona, in the relics of the sky-high Aztec shrines of the Mexican plateaux, or in the igloos of the north coast Eskimoes. These are flashbacks to an earlier stage in the occupation of North America. Written records of the sub-continent tell of a people of many origins: artifacts from before the arrival of European man suggest a similar story.

A PEOPLE WITHOUT A RECORD

There were probably rather more than a million people in North America at the time of the arrival of the Spaniards in the fifteenth century. J. H. Mooney makes this estimate on the basis of maize yields in their presumed oases of cultivation; though others, such as H. J. Spinden, have postulated several times as many. The origin of these inhabitants is even more a matter for speculation than their numbers. It is a riddle which can only be partially answered as archaeologists piece together scattered fragments of information.

Attention has been most consistently directed to the north-west; but it was a Spanish priest, not an archaeologist, who first debated the issue. Joseph de Acosta, sometimes called the Pliny of North America, based his reasoning on the Old Testament:

> All men are descended from Adam [he wrote], so we cannot give another origin to the man of the Indies. . . . All beasts and animals of the land perished except those which were reserved for propagation of their kind in the Ark of Noah . . . [Therefore] for both man and beast, we must of necessity seek a road by which they must have passed from the Old World to the New.

His hypothesis appeared in the 1590's and the idea of a land bridge to Asia was taken up by others. Maps, such as that of Giacomo Gastaldi (1546), joined North America and Asia by a north Pacific land bridge; but it was Russian fur traders and early-eighteenth-century Russian sailors under the Dane, Vitus Bering, who first demonstrated the absence of a link by land. Yet the Strait named after Bering is not a formidable obstacle. It is not much wider than the English Channel, it has many offshore islands, it is traversed by ice bridges in winter and its sea level has varied considerably in time.

If the original inhabitants of America did not enter from the north-west, where might they have come from? They can scarcely have drifted across the narrower parts of the Atlantic Ocean, though a Thor Heyerdahl may yet demonstrate the possibility of this. They might have migrated across the Pacific. The long voyages of the Polynesians testify to their qualities of seamanship, but their historically recorded journeys are of the order of hundreds, not thousands, of miles. Admittedly, there may have been unintentional drifts, but the lie of the land—and islands—in relation to the Pacific wind systems and ocean currents would not have helped. *Kon-tiki* sailed and drifted from east to west. Yet there is eloquent testimony to the diffusion of culture around the Pacific rim. Captain Cook's voyage with H.M. ships *Resolution* and *Discovery* (1776–80) brought to the attention of his company of philosophers similarities between the cultures of the British Columbian Indian and of the New Zealand Maori. The same type of plank house, with its painted corners and lintels, the same designs on blankets and mantles, identical detail in the fish-hooks—all suggested that the coastal Indians had a Pacific origin.

From speculation concerning origins, several fairly incontestable facts may be extracted. First, there is no American evidence of any anthropoid remains to correspond with *Pithecanthropus* or *Neanderthal* man. Secondly, the earliest remains discovered seem to be of inter-glacial origin. Thirdly, there are no known human artifacts beneath glacial deposits; though that is not to say that they might not be found. Finally, the oldest evidence seems to be associated with the western Cordillera.

The north-west route of entry would be at the extremity of this, but in former times the Cordillera in Alaska would have supported a glacial cover of varying extent. The changing distributions of ice during the glacial period are themselves a subject of much speculation, but there are grounds for assuming a land bridge across Bering Strait at the end of the Wisconsin glaciation (*c.* 10,000 B.C.). Before the release of glacial melt-water the sea level stood much lower: it was probably a full 100 feet lower even 20,000 years ago. If men did not migrate across such a narrow water-way, animals did. In any case, early man had boats. The Canadian anthropologist Marius Barbeau quotes the island-to-island migration of the Indians of the Queen Charlotte Islands and their ancestral stories of similar movement as the method which might have been employed.

If palaeo-Indians did arrive from the north-west how might they have penetrated the continent? There must have been one stream of migrants who followed the coastal route; though the icy shores of a glacier-ridden Alaska would have given rise to difficulties. To have followed the invitation of the River Yukon would have led to a virtual *cul-de-sac.* What of the Mackenzie River Valley? It was conceivably free of ice 15,000–20,000 years ago, so that it would have been possible to gain access to the foothill country. Cordilleran Canada offers archaeological testimony which could be used to support this idea.

It is not unlikely, then, that the western mountains became the axis of dispersal, with the waterways of the foothills leading migrants to the east (in mirror image of the migration of Europe's aboriginal population to the peninsular west). Any migrant approaching America from the extreme north-west would find that the way south conducted him to more climatically favoured areas and lands of greater opportunity. It would lead eventually to the constriction of central America where the presence of a high plateau in a tropical setting, the abundance and diversity of plant life favoured the fullest expression of human endeavour. Those who moved into the main body of the North American continent seemed to disperse their energies; those who reached the isthmian plateaux of the Americas found themselves. Farther south, similar high-latitude opportunities were encountered in the Andes. Those who followed the Cordilleran call to its logical conclusion arrived in Tierra del Fuego.

Dispersal resulted in the emergence of different physical features and of different languages (there are a full thousand languages used among the natives of America) as well as the appearance of the more familiar differences in tradition and custom. The development of biological differences takes a long time. This fact actively supports the concept of a succession of migrations perhaps spread over an extended period of time. Eskimo immigration to the Arctic coast of North America is of much more recent origin and there is evidence to support the theory that they arrived in successive waves. It is believed that the last Eskimo immigration to the north-west may have been as recent as a thousand years ago. But to argue by analogy is unwise.

European experiences help but little in the attempt to reconstruct a picture of the primeval peopling of North America. For example, artifacts of the Folsom find in California bear superficial resemblance to those of the Danish Stone Age (c. 3,000–2,000 B.C.), yet other evidence accords to them a position in time parallel to the Solutrian period in Europe (c. 13,000 B.C.). Again, new methods of enquiry, e.g. Indian blood grouping, frequently destroy rather than help to build up existing hypotheses. The Amerind has no sphinx, but he has its riddle just the same.

The next few generations may well witness new answers to the riddle of New World man. Meanwhile, it is evident that geography has its contribution to offer to the solution of the problem. The evolution of early man in the New World can only be appreciated in terms of the evolution of its physical geography.

THE REALM OF THE AMERIND

At the time of the arrival of the Spaniards the natives of North America displayed a variety of cultures. It is debatable how far the word 'civilization' could be applied to their ways of life, though Friedrich Ratzel found a place for it in the human divisions of America which he recognized in his *Anthropogeographie* (1882–91). His classification into the hyperborean, north-western, north-eastern and 'civilized' people of Central America was perhaps an oversimplification. More recent scholars, such as Clark Wissler and A. L. Kroeber, have tried to decide upon criteria for the classification of different native groups. Two simple, yet fundamental, criteria are (i) whether or not the group practises agriculture and (ii) whether their technical and/or cultural achievements are meagrely or strongly characterized.

Several brief examples will serve to illustrate the application of these criteria. The Eskimo 'culture' provided one distinct group: it was (and is) one of lowly technical achievement and no established agriculture was practised. The Indians of the Fraser-Columbia estuary lands fell into a second category. They practised no farming, but they displayed considerable technical achievement and decorative ingenuity. The culture of the plains, with the related eastern forests, was a third type; a little meagre agriculture was practised, but the emphasis was on hunting and trapping. Absence of transport kept the Indian effectively away from the bison. Technical innovation was lowly, though mound building in the Mississippi Valley for protective purposes must not be disregarded. Both plainland and north-west Pacific Indians received some impetus from initial contact with white people. Had not the white advance been so swift and remorseless, a stage of more advanced evolution might have taken place, e.g. among the Iroquois.

Of quite different character were the Indian communities of the south-west. Three principal groups are identified, the Pueblo Indians, the Aztecs and the Mayas. The Pueblo tribes of Utah-Colorado, Arizona and New Mexico were the most modest of the three. All practised agriculture and showed a capacity for technical inventiveness. They reached the height of their development between A.D. 900-1200. In semi-arid farming country they produced fibres, such as cotton, as well as food crops. They cultivated the arts of the potter, the mason and the basket-maker. Their clustered dwellings, sometimes several storeys high, show considerable technical achievement. Their settlements took on legendary characteristics among early explorers—the search for the 'seven fabled cities of Cibola' marked on the Gastaldi map of 1546 continued for a long time.

The Aztec development was the most spectacular and, in its flowering, was recorded by witnesses from the west. City life was supported by a rich domestication of plants—maize and manioc, kidney and lima beans, agave and arrowroot, peanut and cashewnut, sweet potato and artichoke, pineapple and tomato, pepper and prickly pear, gourd and guava, cocoa and cotton, tobacco and coca (for cocaine), yucca and henequin. By contrast, fauna was not pressed into the service of man so readily as flora. The lama (and vicuna), poultry and the bee were domesticated; but until the arrival of the Spaniards there were no domesticated cattle, sheep, horses, donkeys (therefore no mules) or camels for the arid wastes. Nature diminished New World quadrupeds, as Buffon remarked, though she 'cherished the reptile and enlarged the insect tribes'. The Mayan 'civilization' was based upon a sun-worshipping farm community.

Desertion of its cities in the jungle of the limestone peninsula of Yucatan has never been satisfactorily explained. Several climaxes in its pre-Columbian development are postulated. The most fascinating relics are the stone hieroglyphs and calendars which identify the Mayans as a record-keeping people. No one has succeeded in relating the calendars to a precise span of years.

The world of the Amerind, as the native American is commonly called, developed without a beast of burden. Perhaps even more important, it developed without a beast of traction. The wheel and the plough had not the significance of the Old World. The miracle is that in the absence of animal manure a system of arable farming could be sustained, and that in the absence of beasts of burden, urban centres were born and impressive buildings were erected in them. It has been well remarked that if the bison had submitted to domestication a second Thebes might have arisen on the banks of the Mississippi.

THE LANDFALLS OF THE WESTERN WORLD

The most important geographical fact in the history of North America has been its orientation across the Atlantic Ocean to Europe. The Atlantic Ocean is smaller and more negotiable than the Pacific: a richer river network is tributary to it. European man was schooled early in the art of maritime experience. He had two landfalls on the shores of the New World—a northern, springing from a Scandinavian source, and a southern, springing from a Mediterranean source.

The northern landfall belongs to the first years of the eleventh century. For the Norsemen, the stepping-stone route by way of Iceland and the Greenland settlements resulted in the voyage of Leif Ericsson. The early-eleventh-century *Codex vidobonensis* records his discovery of *Helluland* (or stony land), *Markland* (or woodland) and eventually *Vinland* (the land of the vines). An east-coast thesis equates the three names with Labrador, Newfoundland and Cape Cod area respectively; a Hudson Bay thesis equates them with Hudson Strait, James Bay and southern Ontario. There is much fiction and rather less sober fact about the continuity of the Scandinavian contact with the New World in the middle ages. Did the Mediterranean mariners who set out nearly 500 years after Leif Ericsson by way of a southern route hear of western landfalls from Scandinavian fishermen?

If they did, that was only one fact which encouraged Atlantic adventuring by the Mediterraneans. The Azores had been rediscovered by Prince

Henry the Navigator in 1431. The intellectual curiosity of the age, given patronage by monarchy, encouraged exploration. Spain, unified against a common Moorish foe, turned its energies and mobilized forces overseas when released from African invasion. New Spain preceded New France and New Britain in America. The Aztec realm was conquered in the 1520's. Over a century before the *Mayflower* came to New England, Cortez and his adventurers entered Mexico City. The chronicles of Spanish priests recast in the mould of W. H. Prescott's *History of the Conquest of Mexico* make a story as exciting and exotic as almost anything written by Rider Haggard.

The economy of New Spain was based upon the theory that a colony exists for the benefit of the home country. By 1600, with roughly 5,000,000 Indians under the control of some 200,000 Spaniards, the land was divided into great feudal estates; while the extended Spanish empire was administered from a mainland captaincy-general in Guatemala and an island captaincy-general in Havana. Pastoral pursuits were added to indigenous agriculture (Columbus had already introduced horses, mules, cattle, sheep, goats and hogs on his second voyage of 1493), mining was spurred forward (St. Luis Potosi became the biggest city in the New World), industry lagged behind. African slaves were introduced to make a trinity of human elements in Central America and Caribbea. Side by side with the system of exploitation arose a colourful Catholic civilization with churches, monasteries and schools.

Another Catholic realm was established to the north-east. New France was born on the banks of the St. Lawrence in the early seventeenth century. From the magnificent citadel of Quebec the arc of empire extended inland. As conceived by La Salle, France had a vision of a French continent. Within a century of the settlement of Quebec (1608), French influence had penetrated a full 2,000 miles into the Laurentian hinterland. The mobile French seemed to follow headlong wherever the waterways led, looking for an Eldorado and seeking the Indian 'Big Sea Water'. The result was a large, loosely knit domain held together with the greatest difficulty. It eventually stretched from the seasonally frozen watergate of the St. Lawrence, with its tightly packed farms running from river-line to backwoods, to the watergate of the Mississippi, with its sub-tropical port of New Orleans. Soldier, priest and trapper, rather than settled farmer, held together the intermediary key points—Niagara, Detroit, St. Marie, Fort Miami (on the Maumee River), Vincennes (on the Wabash River), Chartres (on the Mississippi opposite St. Louis). It was a colonial empire which suffered a kind of 'sweet water' dropsy. French energies seemed to be dissipated in space—space which was inferior as a settlement area to most of France itself.

British colonial settlements on the Atlantic seaboard between those of New France and New Spain's Florida were hemmed in between the mountains and the sea. They were of several types and, initially, they stood in rivalry between those of the Dutch and the Swedes, which they eventually absorbed. New Holland, having some 10,000 settlers around the lower Hudson and centred on New Amsterdam (subsequently New York), became British in 1604. A legacy of Dutch place-names recalls its former extent. New Sweden, on the banks of the Delaware, was never such a vigorous colony.

There were three principal British settlements—the northern colonies of New England, the plantation settlements of the south-east (Virginia, the Carolinas and Georgia), and the intermediary tidewater lands (Pennsylvania, Maryland, Delaware, New Jersey). The Atlantic settlements had no unity of origin. Some were lands of refuge, granted their charter of right by the king: some were proprietary colonies sponsored by trading companies. Those who emigrated to them had modest qualifications as pioneers. They entered the territory of the eastern woodland Indians, whose elementary husbandry taught them but little (e.g. tree girdling for land clearance, maize cultivation). For all immigrants, it was arrival in a land with very different climates—'a lustier land, with hotter sun and more tempestuous rain', as Carl Sauer has characterized it.

Captain John Smith, 'Admiral of New England', was among those who painted a bright picture for the dissenters and nonconformists who sought a new home in New England. 'Of all four parts of the World I have yet seen not inhabited . . . I would rather live here than anywhere.' But philosophy alone could not convert the free timber, stone, slate, iron ore and fish into useful commodities. William Wood's *New England's Prospect* (1634) expressed a need for 'ingenious carpenters, cunning joyners, handie coopers, good bricklayers and tylers'. As craftsmen appeared, clapboard houses succeeded log cabins and Wren-style churches in wood lifted their slender white spires above the tree-tops. Early settlements were nucleated. William Bradford in his *History of the Plymouth Plantation* (1603–48) told how 'on Tuesday, 9th January, 1621 . . . we went to labour in the building of our towne, in two rows of houses for the more safety'. Protection of beast as well as of man was an essential need. Beyond the garden and the fenced homefields extended the outlands with forest and salt-marsh grazings, fishing and shore rights. Wethersfield, Connecticut, was such a settlement and its 1641 plan is reproduced in the *Atlas of Historical Geography* (Plate 41D). The Puritan settlers may have been neither farmers nor handymen by training, but they had two inherent attributes as colonists. They were workers who put into practice the Parables of the Vineyard and the Talents; and they were individualists who were able to withstand the bane of solitude. Communities of such men and women gave rise to the six small states of New Hampshire, Maine, Vermont, Massachusetts, Connecticut and Rhode Island between the White and Taconic mountains and the sea with its offshore banks.

The southern colonists were different. In 1584 Hackluyt had already proposed New World forts from which to attack the King of Spain's lands so that 'he shall be as bare as Aesop's proud crows'. In 1607 the Jamestown colony was established by the Virginia company. The colonists were principally ordinary artisans and merchants, rarely cavaliers as propagated so widely by W. M. Thackeray's *Virginians*. The wealth of the South was not quickly realized; it awaited the introduction and establishment of staple crops—a tale told in L. C. Gray's memorable *History of Agriculture in the Southern U.S.A. to 1860* (1933). Physical as well as mental climate inhibited early development. William Byrd in his *History of the Dividing Line* (1685) pointed this out:

> Surely, there is no place in the world where the inhabitants live with less labour than in North Carolina. It approaches nearer to the description of Lubberland than any other by the great felicity of the climate, the easiness of raising provisions and the slothfulness of the people.

It was even better if others could work 'in the heat of the sun' instead of the white man. Small wonder that the institution of slavery found a ready home here. The great days of slavery date from the end of the seventeenth century, when for a good many years fully 20,000 slaves were imported annually. By 1750 a state such as South Carolina had twice as many Negroes as white inhabitants. Given slavery and southern staples such as cotton, tobacco and rice, the plantation evolved and that lucrative triangular trade which knitted together Great Britain, West Africa and the south-eastern states.

Between the New England settlements and the southern colonists were those of the 'Tidewater Lands'. In some respects, the Delaware and Chesapeake bays, with their rich and diverse hinterlands, were the most important British landfalls. In general, they were settled later. More frequently, the settlers were of farming stock—from Hanoverian Germany as well as Hanoverian England. In the tidewater lands met and mingled a variety of agricultural crops, stocks and techniques—both native British and continental European. The tenant or *petit propriétaire* of Europe became the substantial owner-farmer when translated to the middle Atlantic states. This was the birthplace of so much of America's mixed husbandry. Northern and southern farming met here. It was also a hearth of Quakerdom, with Philadelphia, the city of brotherly love, as its focus. In a land of so much wealth and opportunity, the Quaker became rich in spite of himself. It was here that the Declaration of Independence was signed and in the same Atlantic foreland the final capital of the federal state was located.

The English seaboard settlements, with their complementary products born of a diversity of environments, were linked by sea routes rather than land routes. Bridging the many broad and braided rivers with their capricious flow was a problem even after colonial times. Turnpikes, as shown on the Post Road Map of 1774, might attempt to strengthen land links, but it still took three times as long to travel over them as to journey by sea. From Boston to Savannah, for example, occupied fifteen days overland and five by sea. Colonial America not merely looked to the sea, but was bound together by it.

Yet before the time of independence the seaboard state was already looking at what Thomas Jefferson called its landboard. Inland penetration was not easy. Watercourses with majestic names might open up imposing valleys, but none save the exceptional Hudson invited navigation. Rapids and shallows gave rise to hazards and portage. Often, the humble riverbank trail was a more important link than the waterway itself. Woodlands also obstructed movement—in the kindlier climates of central and southern Appalachia blanketing the crests of ranges 5,000–6,000 foot high. On his fine map of 1755 Lewis Evans identified the Ouasioto Mountains of Virginia, and wrote across them: 'A vein of mountains about 30 or 40 miles right across, through which there is not yet any occupied path in these parts.' It is not surprising that British settlement had barely penetrated 200 miles inland when the thirteen states declared their independence in 1775. The Appalachian piedmont was largely frontier country. It was the American's first taste of the west, with a Darwinian struggle for existence against natives as well as nature. The frontiersmen needed freedom to manœuvre and resented the authority of the coastal administrators to whom they were yoked. The obstruction of Appalachia—or the concept of the obstruction—was of primary consequence for English-speaking Americans: it kept them compact upon the coastlands of what Constantin Volney called 'the Atlantic Country'.

But if settlement was compact it was not crowded. There was a high birth rate, as befitted the need of the land. 'As soon as a person is old enough', wrote Pehr Kalm in his *American Journey* (1748-50), 'he may marry without fear of poverty, for there is such a tract of good ground yet uncultivated.' In Europe, Thomas Jefferson observed a generation later, the object was to make the most of land, labour being abundant; in America it was to make the most of labour, land being abundant. In the 1770's there were 3,000,000 inhabitants in the emerging confederacy. Despite immigration, perhaps two thirds of them were the descendants of the original settlers. With England a quarter of a year away it is not surprising that they thought as Americans. 'Seas roll and months pass between the order and the execution', complained Edmund Burke of trans-Atlantic negotiations. The reasons for the political break of the colonies were human as well as economic. The years 1775-6 are very much a watershed in the story of the New World. British authority was cast off by an organized federation and loyal elements made their retreat to the north. Canada was born—or rather reborn—in the process which brought the Empire Loyalists to the Maritimes, to Quebec's eastern townships and to the north shore of Lake Ontario. Here they became a buffer to resist American penetration and a balance to the strongly entrenched *Canadien*. The United States of America was conceived between the sea-mark and the mountain, as Stephen Vincent Benet put it, but the birth of the American awaited the crossing of the range. The crossing was not long delayed. Oceanic imperialism in the east was rejected and forgotten and the United States launched its own imperial programme in the west—the west beyond the west of the Appalachian piedmont.

THE POLITICAL DEFINITION OF NORTH AMERICA

The definition of the U.S.A. is largely a study of the elimination of the Old World from the New. It involved the establishment of an 'Anglo-Saxon' border in the north and a 'Latin' boundary in the south. The creation of its northern boundary involved a compromise with Great Britain, where the Canadian boundary was agreed in sections. The settlement of 3rd September 1783, by which the independence of the U.S.A. was formally recognized, advanced the frontier westwards to march beside that of New Spain (Figure 13). Simultaneously, the Great Lakes system became the northern boundary, with the narrows as fortified strong points. The Niagara peninsula became a minor battleground in the War of 1812, and its creeks (or rivulets) still bear names which relate them in a strategic system of defence lines to the Niagara River. The second stage in the definition of the northern boundary was taken in

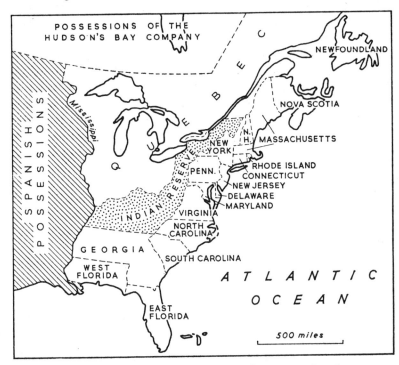

Figure 13 *Political Boundaries on the Eve of American Independence*

European wrestled with Indian as well as European over the definition of American boundaries. The map summarizes the situation just before the thirteen colonies declared their independence. The boundaries of the colonies are also defined

1818, when a 'Prairie' frontier convention agreed upon the 49th parallel (with adjustments along the Lake of the Woods boundary in 1842). The third stage concerned the maritime extremities. The Webster-Ashburton Treaty fixed the detail of the Maine-New Brunswick boundary on 9th August 1842, while the Buchanan-Pakenham Settlement brought peace out of near hostilities on the Pacific coast on 15th June 1846.

The 'Latin' boundary resulted from an interplay with France, Spain and (after 1821) the independent realm of Mexico. In 1800, by secret treaty, Spain surrendered Louisiana and New Orleans to a France which still cherished visions of restoring its North American empire. After the renewed outbreak of hostilities between Great Britain and France (1803), Thomas Jefferson visited Paris and purchased Louisiana for the U.S.A. The purchased land had no formal southern boundaries until 1819 when east Florida was bought from Spain. The Mexican declaration of independence (1821) was guaranteed to raise eventual problems; partly because of the uncertainty over the Sabine River and partly because of the steady advance of American settlement into the no-man's-land between Louisiana and the Rio Grande. Texas became a frontier tract of Mexico, settled principally by Americans, who in 1836 declared their territories an independent republic. In 1845 the Republic of Texas was annexed by the U.S.A. and in the peace settlement following the ensuing Mexican Wars (1846–8) the Rio Grande was accepted as the new boundary. New Mexico and California thereby became a part of the U.S.A. and the frontier formally shifted to the Pacific shore. Apart from minor adjustments (e.g. the Gadsden Purchase of 1854), the Mexican frontier has been stable. Indeed, the boundaries of mainland U.S.A. have now been stable for more than a century.

The 'Anglo-Saxon' and 'Latin' frontiers were defined; but an unexpected 'Slavonic' enclave remained. Russian Hill in San Francisco is a place-name which recalls the depth of Russia's penetration along the Pacific coast. The Oregon settlement eliminated Russia south of the 49th parallel, but a governor was settled regularly in Sitka to keep an eye on the Alaskan fur trade. In 1867 Secretary of State Seward bought Alaska from Russia for the sum of $7,000,000. In agreeing upon its Canadian limits, a long coastal 'panhandle' (based upon an Anglo-Russian agreement of 1825) was extended southwards towards the Skeena. The America-Russian boundary was drawn through the Bering Strait. A direct consequence of the process of definition was the successive acquisitions of Public Lands by the central government. The growth and disposal of the Public Lands reflect the peopling of the west. Extensive Public Lands remain under the control of the federal authority in western states such as Nevada, Utah, Colorado and Idaho. Problems of absentee control are rooted in them.

High-latitude definition of North American political boundaries had little significance in earlier times; but Canada's arctic limits were destined to acquire a new importance in the twentieth century. It was a Canadian,

Poirier, who put forward the sector principle of sovereignty and in 1909 Canada claimed a triangle of island territories subtended by the broad base of the Arctic coast. A tenth province entered the Canadian federation in 1953 when Newfoundland opted for incorporation. The crystallization of the political frontiers of mainland North America was complete.

THE FRONTIER IN RETREAT

Throughout the period when the political frontiers were crystallizing most Americans were probably moved by the 'frontier' in quite a different context. The frontier of settlement meant more than the frontier as a political boundary. The concept of the frontier of settlement has given rise to a vast and stimulating literature. If the 'Wild West' stirs the imagination of most schoolboys, the process of its taming encourages the enquiry of scholars. The frontier and the 'West' have been in some respects synonymous; though a succession of 'Wests' rather than a single 'West' must be identified (Figure 14). The West which was the 'Old West' yielded to the West which became the Middle West: the Far West gave place to the West which was perhaps the true West—the High Plains and foothills from Montana, through Wyoming and Colorado, to Texas.

Occupation was no simple matter. Lines may be drawn upon a map to illustrate the advance (Figure 15), but they can only be very rough generalizations. First, there were many enclaves of unoccupied agricultural land behind the frontier of settlement. There were certainly plenty of exclaves beyond it, such as the farming colonies of the Mormons of Utah or of the Arkansas settlers in Fenimore Cooper's novel *The Prairie*. Secondly, there was the irregular succession of migrations resulting from mineral discoveries—the gold rushes (to California, Dakota, Yukon, British Columbia and Laurentia), the silver rushes (to Nevada or Colorado), the oil rushes (to Texas and Oklahoma). The Californian gold rush was in some respects the most unique of these episodes. Though there was simultaneous agrarian occupation of Oregon to the north, the peopling of the Pacific coastlands was spurred forward much more energetically by chance of swift gain. The three-month hazard of the overland trail to California or Oregon (described in Francis Parkman's clear prose), the route round the Horn followed by clippers such as that taken by Henry Dana during his *Two Years before the Mast* (1840), and the isthmian route over Panama brought substantial settlement to the Pacific slope before most of the High Plains and the intermontane basins were

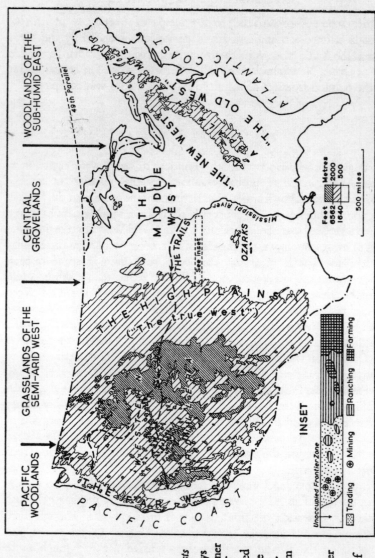

Figure 14 *The Succession of Wests*

The map portrays in a general manner the succession of wests as conceived in the mind's eye of the American. The inset diagram summarizes a commonly prevailing frontier situation during the occupation of the Great Plains

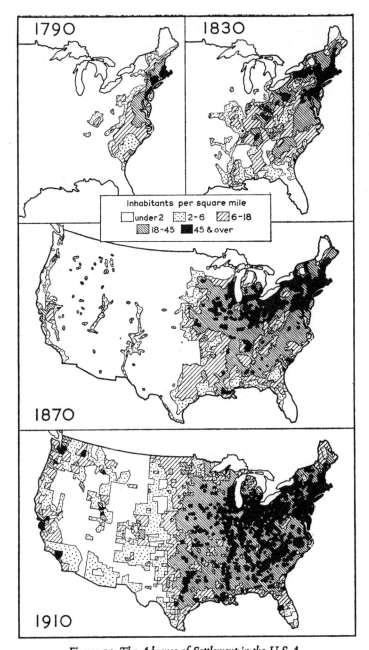

Figure 15 *The Advance of Settlement in the U.S.A.*

The expansion and concentration of settlement in the U.S.A.,
based primarily on the censuses between the first and twelfth.
(With modifications, from C. O. Paullin, *Atlas of the Historical
Geography of the U.S.A.*, Washington, 1932)

permanently occupied. Even after the census of 1891 had declared the 'frontier' closed, there was plenty of open land. The nineteenth century, encompassing most of the epic of colonization, may with justification be called the century of the New World in itself.

Once the 1783 agreement was concluded, the United States began western expansion in earnest. First came the settlement of the Ohio lands. Americans were well aware of their value and Lewis Evans, on his map of 1755, had tried to impress upon the disregarding English that they were 'as great a prize as has ever been contended for'. A rich literature of schemes for plantations on the Ohio is to be found. Expansion of Americans implied a corresponding contraction of Indians. Tribal tracts such as those of the colourful Cayugas, Onondagas and Miamis slowly disappeared from the face of the land. The Indian population was partly subordinated by force of arms, partly by exchange of territory and formal allocation of reservations (cf. Paullin, Plate 36). *The Last of the Mohicans* was laid to rest by Fenimore Cooper and the Indian soon became fiction rather than fact. Yet the persistence of an effective Indian resistance was prolonged in the plain-lands. Records of the Oregon Trail recall the organization of para-military convoys required to pass through Indian country. It is still exciting to encounter files of country newspapers (such as the *Litchfield Gazette* of Minnesota) where the compositor was setting up print while Indian raiders were being sniped from the windows of the press room or to remember that the last battles in the west, where General Custer's men were dying 'with their boots on', were fought within living memory (1877-9). It is as hard to realize the recency of permanent white occupation of the American countryside in time as it is to appreciate the perspectives of the New World in space.

The course of empire which carried settlers westward transformed their attitude to the motherland of the Atlantic seaboard. Constantin Volney observed in his *View of the Climate and Soil of the U.S.A.* (London, 1804):

> The inhabitants of the Atlantic coast call the whole of this the Back Country, thus denoting their moral aspect—constantly turned towards Europe, the cradle and focus of their interests. It was a singular though natural circumstance that I had scarcely crossed the Alleghanies before I heard the borderers . . . of the Ohio give, in their turn, the name of Back Country to the Atlantic coast, which shows that their geographical situation has given their views and interests a new turn.

The relative importance of the 'borderers' was modest, for in 1790, out of America's population of 4,000,000 not much more than 5 per cent lived beyond the range. But numbers were to grow rapidly.

Beyond the Mississippi the name Louisiana was writ large upon the maps of the day. Explorers were soon pushing through its tracts. Some were sent by the U.S. Army. Lewis and Clark undertook incredible hazards on their expedition to the Pacific north-west by way of the Missouri (1804–6). Other pathfinders were sponsored by traders. Among them were the so-called Astorians (1811–13) who sought to establish John Jacob Astor's fur trade. Jedidiah Smith explored the Great Basin region (1826–7); Frémont pushed along the flanks of the Sierra Nevada (1845–6).

The first stages of colonization relied upon the horse, the ox, 'Shanks's pony', the canoe and the flat boat. The Conestoga wagon, with its varied designs, had its origins on the Atlantic seaboard, though the fullest expression of wagoning is associated with the billowing canvases of the prairie schooners. The horses and oxen which served as draught beasts at first found adequate fodder on the natural grasslands of the Middle West. Commercial farming, looking to grain and especially to corn, could only succeed subsistence farming if the products of the soil could be transformed into goods valuable in proportion to bulk, if a new form of transport became available, or if they could carry themselves to market. Corn could be converted into brandy and withstand transport costs or into meat which walked to market. The droves of cattle, swine (hogs to Americans) and even turkeys which migrated from Appalachia and the upper reaches of the Ohio to converge on the consuming centres of the tidewater lands were New World counterparts of the more modest migrations to London from raising or fattening centres such as Wales and East Anglia. Beasts might move in droves of several thousand head and at 8 or 9 miles a day. Migrations probably reached their peak in the days when Mrs. Trollope was scandalizing the *Domestic Manners of the Americans* (1832) and Charles Dickens was writing his *American Notes* (1842).

Movement was aided by both natural and artificial waterways. The Great Lakes offered navigable waterways: so did most of the eastern tributaries of the Mississippi. Navigable waterways were a new feature to Americans from the Atlantic coastlands. The Erie Canal was the first attempt to create a continuous waterway between the east coast and the interior. Soon afterwards, the first Welland Canal provided contact between lakes Ontario and Erie, and the short-lived Erie-Wabash Canal struck across the Mississippi watershed. The passage of settlers was eased especially by the Ohio which had as much right to be called *la belle rivière* by the republican Americans as it had been by the royalist cartographer Delisle. Of all the rivers, the Mississippi stirs the imagination most profoundly; though at the same time it wrinkles the brows of those responsible for its control. The Mississippi was for a brief while the great north-south artery of the U.S.A. By the late 1830's steamboats were already operating on the Great Lakes and were replacing flat boats on the rivers in the heyday of the stern-wheelers (1840–50). For Mark Twain *Life on the Mississippi* as a river pilot was a discipline as thorough and instructive as that received in any university.

The 'father of rivers'—so Indians called the Mississippi—also had children who were not to be despised. Graham Hutton has observed that the Illinois River at Peoria is as wide as the Danube at Budapest. The magnitude of the right-bank tributaries of the Mississippi is even greater; but they are moodier. Valleys such as those of the North Platte and the South Platte, the Canadian and the Red rivers, might open up trails through the High Plains and passes through the American Rockies, but shallows and obstructions prohibited more than very limited navigation.

The role of rivers and canals was rapidly subordinated with the application of steam power to land transport. The Appalachians were crossed by a series of railways in the 1850's and the spread of the mid-western network occurred in the 1860's. The first trans-continental line (Union Pacific and Central Pacific) was aided by government loans. These reflected the terrain through which the line passed—$16,000 a mile in level country, $32,000 a mile in the foothills, $48,000 a mile in mountain country. The golden spike which united its last rail links at Promontory Point in Utah in 1869 may still be seen in San Francisco. By this time, however, overland mail routes, by coach and courier such as the celebrated Pony Express, ran from plainsland railheads to the West coast. The railroad age of the U.S.A. was as fevered as that experienced in any European land. From 30,000 miles on the eve of the Civil War the length of line increased to 167,000 miles by 1900. Railways were initiated privately and America's railways today (their names telling tales of historic amalgamation) remain in private hands. Striking transversely across the Mississippi route, the railways swiftly altered the downstream course of commodity movement. Flow lines assumed an east-west direction, to and from the centre of industry, consumption and export in the north-east. At the same time Federal Land Grants placed extensive areas in the control of the railroad companies and prospective pioneers were able to buy 'land exploration tickets' to the Golden West.

Occupation of the land was organized against the background of formal survey of township and range (cf. p. 25), of military reserve and grants to land companies. Later, the Homestead Act (1862) eased the acquisition of land. Major administrative units had also to be created. These called for reference back to parent states on the Atlantic seaboard and their birth was rarely achieved without stress and strain. States were eventually born out of 'territories' and 'territories' changed shape substantially in the process. Ohio was the first state to be born beyond the range (1803): Indiana territory changed shape three times (1803-5, 1805-9, 1809-16) before achieving stable boundaries. Beyond the 'territories', which were usually subdivided after occupation (cf. Michigan territory, which embraced the later Wisconsin), lay 'unorganized' lands. Photographs were taken of the 50,000 settlers who awaited the signal for the opening of Oklahoma at twelve noon on 22nd April 1889. Arizona and New Mexico achieved statehood as recently as 1912. Once states were born, they had to be christened. The lyrical quality of their names is now lost in familiarity. Sometimes, it is interesting to reflect upon state names which were rejected: Transylvania, Charlottania, Vandalia.

The selection of a settlement site (given a range of choice) and the method of colonization varied widely. Sometimes the settler in the woodland zone, having acquired familiarity with the flora, used it as a key to the detection of good soils. Thomas Hutchins had pointed out this method in his field notebooks from the Ohio lands. E. A. Talbot, author of *Five Years' Residence in the Canadas* (1820), observed that 'land upon which black and white walnut, chestnut, hickory and bass-wood grow is esteemed the best'. Observations on flora by Lorin Blodget in his *Climatology of the U.S.A.* (1857) suggested affinity of agricultural opportunity from the Red River Valley to the Mackenzie's southern tributaries. Some digested the wisdom of emigrant literature such as J. M. Peck's *Guide for Emigrants to the West* (1837) before they set out. Of those who engaged in land-breaking, quite a few were authors in their own right. O. E. Rølvaag's *Giants in the Earth* (London, 1927) has as its background the taming of a tract of Dakota grassland. There were some women who held their own in the masculine world of the frontier. Susanna Moodie's *Roughing It in the Bush* (London 1854), records the detail of wild lands in southern Ontario which were broken by her family.

There were professional land-breakers north and south of the border. A contemporary document described some of the American professionals as 'playing at leapfrog with their lands'. 'So soon as they have cultivated a spot that any newcomer likes, they sell it and move higher up country . . .

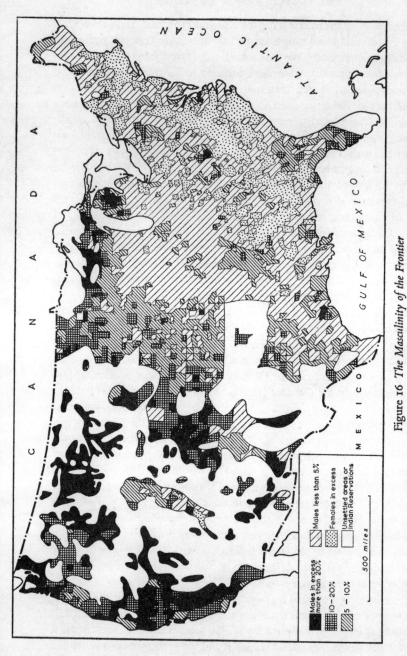

Figure 16 *The Masculinity of the Frontier*

The pioneer fringe had an essentially masculine character. This map illustrates the distribution of the predominant sex according to the U.S. Census of 1890–1. It derives from the first formal *Statistical Atlas of the U.S.A.* (Washington, 1898) edited and compiled by Henry Gannett, 'geographer of the tenth census'

Legend:
- Males in excess more than 20%
- 10–20%
- 5–10%
- Males less than 5%
- Females in excess
- Unsettled areas or Indian Reservations

500 miles

CANADA

ATLANTIC OCEAN

GULF OF MEXICO

MEXICO

[perhaps] four times in the space of ten years.' Canada had 'men who throve on clearance as a profession', according to a report on Bruce County in the *Transactions of the Ontario Board of Agriculture*. They occupied 'a new farm every five or six years' and then 'moved on'. Elsewhere, these colonists were graphically described as 'hating the plough'.

The peopling of the western lands was initially undertaken by men rather than women, as Figure 16 shows; young men, in particular, as is evident from Figure 17. It was also undertaken by immigrants of different origins at different times, and by different racial groups in different types

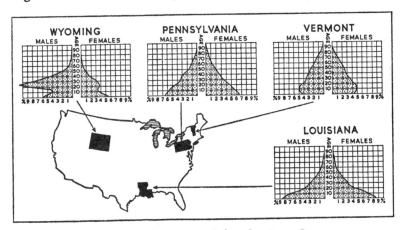

Figure 17 *Population Pyramids from the 1890–1 Census*

The four pyramids illustrate contrasts in the structure of population in four states during a critical period of their evolution. The states differ in the nature of their physical background and in the degree of their economic evolution

of country. First immigrant records were kept in the 1820's, and from the later censuses it is possible to trace the shifting distribution of foreign-born white population. This was originally most pronounced on the frontiers of settlement. Subsequently, it became most marked in the older cities of the north-eastern seaboard and (as might be expected) along the Canadian frontier. Among other ethnographic variations, religious groupings have shown well-defined distributions. The persistence of the Baptist faith in the Negro regions of the south-east, the stronghold of Lutheranism in the northern parts of the Middle West, the Catholic preponderance in the urban areas of the north-east provide illustrations. In occupying new land men were forced to make adjustments to it. The peopling of the west is therefore a study in the change of old institutions as well as the growth of new ones.

The chronology of settlement of different racial origins is seen in both large-scale and local distributions. Looking at the national scene, the Middle West was peopled predominantly by Americans, though the tide of west European immigration was flowing before it was completely settled. Irish immigration reached a peak soon after 1850: German and Scandinavian between 1870 and 1900. Minnesota is a 'Scandinavian' state, with Danes, Norwegians and Swedes concentrated in the centre and south, and Finns generously distributed in the coniferous north. States such as Wisconsin and Michigan have pronounced German concentrations; while density of Swiss settlement has even given rise to the local nick-name 'Swissconsin'. By the time that the High Plains were being penetrated by the lines of steel, improved transport in Europe and across the Atlantic was adding a large East European component to the stream of immigrants. An oriental stream also began to flow to the Pacific coast. It is estimated that between 1820 and 1900, 17,000,000 aliens entered the U.S.A. (Figure 18). Italy's contribution, reaching its peak between 1900 and 1920, moved principally to the towns: coloured peoples from Caribbea (such as Porto Ricans) have subsequently crowded into urban areas.

Local distributional patterns tended to reflect the chronology of arrival because the first settlers commonly occupied the better-drained lands or the sites most accessible to the original lines of communication. The Middle Western states provide good examples of this rough-and-ready law in operation, for they have a rather unexpected diversity of soils in close juxtaposition. In the scramble for lands and in the trial and error of farm practices which succeeded it, much land was transformed which might have been better left in its natural condition. Earlier settlement in the Appalachian states of New England and New York illustrate this no less than later settlement in the semi-arid realm of the south-west. Farm abandonment and the retreat of the frontiers of cultivation are a feature of the north-east, while the seeds of tax delinquency matured even earlier on much land in Oklahoma, Texas and Kansas.

Part of the difficulty was that movement west carried settlers to unfamiliar environments. The familiar grove country of big woods and taller prairie grass yielded slowly to shorter grasses beyond the Mississippi-Campestria, as it was identified regionally. Long tongues of woodland penetrated the river valleys and great islands of woodland clothed the Ozark-Ouachita uplands; but between the valleys the open landscapes of

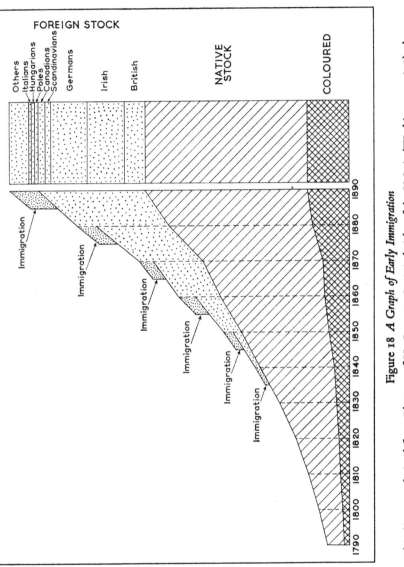

Figure 18 *A Graph of Early Immigration*

This diagram, derived from Plate 22 of H. Gannett, *Statistical Atlas of the U.S.A.* (Washington, 1898), shows the comparative growth of the different elements in the population of the U.S.A. down to 1891

the west replaced the closed landscapes of the east. 'Not a stick of timber except a few clumps of cotton wood', the explorer Zebulon Pike recorded of the lands beyond the Arkansas Valley. Other cartographic wanderers in the wasteland inscribed the word 'Zahara' across the treeless country, and mapmakers persistently copied it. The word 'Badlands' was speedily applied to the whole of the semi-arid area, and people passed through its lunar landscapes on a long trail to a greener west. It was not only low rainfall, but also absence of surface water, which reduced the appeal of the plainlands. Drought of some kind was certain: men learned to gamble with the degree of drought.

The first settlers of the high plains were pastoralists, who established in it their grazing kingdom. It was 'high, wide and handsome' country and its magnificent distances exhilarated men until they were awakened to its deceptions. It was a country to breed a new generation of prophets. Among them stalked William Gilpin, first Governor of Colorado, re-christening the arid plains 'the pastoral garden of the world' and locating the future centre of control of America in the Mississippi Valley.

Geography does not always determine in the way that men expect. Within a matter of months of the publication of Gilpin's thesis the U.S.A. seemed to be as profoundly divided latitudinally as Gilpin had forecast that it would be united longitudinally. If the American Civil War was not born of geographical causes, its background of secession and slavery was partly rooted in them. The southern states, or the Confederacy which ran from Arkansas through Tennessee to Virginia, sought to secede from the Union.

Among the many consequences of the war were the emancipation of the Negroes (1865), the economic exhaustion of the southern states and the diversion from them of the ensuing stream of immigration. Measured by the criteria of birthplace and parentage statistics, the south-eastern states became one of the most American parts of America. Their subordination was exaggerated by the opening of the lands beyond the Mississippi. They were not merely by-passed by those who went west, but ignored by an immigrant body which neither knew of their geographical circumstances nor cared for their historical antecedents.

The retreat of the frontier was not only speeded by improved communications and the increase of immigrants. New technical equipment aided the advance. The steel mills of the north-east specialized in the manufacture of farm implements. Pittsburgh had an output of 30,000 ploughs p.a. in the 1830's. After the Civil War the arsenals of the Union were able literally to beat their swords (and cannons) into ploughshares.

Moreover, with improved and cheaper methods, they were able to manufacture implements such as the 'chilled steel' plough which were more resistant to the deep and tough-rooted grasses of the west. At the same time cheap steel was basic to other inventions which eased the occupation and development of the western lands. In the semi-arid west the railroad companies were the first to put down deep bores and to erect windmills to pump their water. Farmers followed suit. Barbed wire enabled animal control and the possibility of mixed husbandry in the midst of the cattle kingdom. New techniques not merely meant easier control of the farmed land, but possible occupation of land which with more rudimentary equipment was beyond control. The application of mechanical energy to farming represented the last stage in the process of advance. 'Machines for abridging human labour are especially desired in America', Dr. Collin wrote in 1789; in the 1850's Sir James Caird observed incipient steam ploughs in the Middle West. Immigrants no longer wrote their epitaphs in the furrows of the fields.

The establishment of growing order in the disorderly world of the settler can be traced in many ways. The reports of the Commissioners of Agriculture provide one source. They not merely reviewed the character of land use in particular areas, but also made early efforts to improve conditions. The potentialities and shortcomings of specific tracts of land might be the object of their survey. Sometimes, materials were presented cartographically. The Geological Survey also made its contribution. The more mature settlements, spurred by local rivalries, took pride in producing their own county atlases. Several thousand different atlases and plat books were published between 1850 and 1920. They provide sample illustrations of the order and organization brought into a county within the span of as little as a generation. The *Atlas of Litchfield County, Minnesota* (1878) is one example. Canadian counties followed the American pattern, e.g. *Illustrated Historical Atlas of the County of Halton, Ontario* by J. H. Pope, Toronto, 1877, which also illustrates the fate of the farmland first colonized by the 'yeomen' of Nelson (cf. p. 26). By the end of the nineteenth century the evolution of the U.S.A. was being presented in national atlases. The passage of 60 years has converted Henry Gannett's *Statistical Atlas of the U.S.A.* (1898) into an historical document. It was among the first to make detailed use of census material, specifically that of the eleventh census. It was, moreover, published at a particularly significant time. Ethnographically, the U.S.A. had become Israel Zangwill's 'melting pot'; industrially and agriculturally, it had assumed world leadership.

One of the most remarkable features of the American scene in the latter half of the nineteenth century was the multiplication and growth of urban centres. In 1900 the population of the U.S.A. was 76,000,000. A third of this total were city-dwellers. Indeed, the population resident in cities of more than 8,000 grew from 5,000,000 in 1860 to over 25,000,000 in 1900. The peak period of immigration, with immigrants exceeding a million annually during 6 years between 1903 and 1914, exaggerated city expansion. Chicago was the apotheosis of this growth: it claimed a million residents in 1900. Natural increase and immigration were so great that the western lands could be settled simultaneously with the growth of the cities. In the earlier stages the effects of urbanization were most pronounced in the north-east. They had their complement in the depopulation of rural areas. Middle West townships already showed a decline in the 1880's, but nothing like that experienced in the rural areas of New England.

The rise of the towns was a reflection of the growth of industry. As in the Old World, America's early industry looked to wind and water power for energy; but it was of modest importance during the colonial period when industry was discouraged by the mother country. New England watercourses such as the Merrimack provided ideal settings for workshop during the first phase of development. The rise of steam power called for fuel other than the lumber of the Atlantic woodlands. This was to hand in Appalachian coal, for the resources of Pennsylvania and West Virginia were known long before their commercial exploitation. However, they had little value until they could be transported—and the railways, deriving their own energy from coal, provided the answer. Almost at the same time as the railroads were penetrating into the Appalachian coalfields, new methods of producing cheap steel were being introduced. Industry, born on the Atlantic seaboard and partly transformed *in situ*, began to move towards the coal. The biggest single industrial concentration grew up on the bituminous fields around Pittsburgh (Fort Pitt in the colonial days), while a complex of other industrial towns spread along the valleyways of the Ohio, Monongahela and Youhiogheny. The growth of Pittsburgh and its related iron and steel centres was accelerated by the Civil War. Between 1860 and 1880 the total of iron- and steel-manufacturing plants grew from 402 to a full thousand.

The centre of America's industry shifted to the coalfields, to the lines of communication which radiated from them and towards sources of raw material, such as iron ore, which were complementary to them. Yet industry in the east-coast cities was sufficiently entrenched to attract fuel and raw materials; while it had the pick of skilled as well as semi-skilled factory labour which began to emigrate from western Europe in the later nineteenth century. A large percentage of the Europeans who came to the New World were townspeople and were willingly absorbed in the eastern cities of the U.S.A. Change in structure as well as growth characterized the industry of the Atlantic seaboard; with diversification of light industry generally taking precedence over the heavy industries which were based on New York, Philadelphia and Baltimore. The Pittsburgh area therefore evolved less as a rival to the seaboard cities than as a complement to them. By the end of the Civil War it was the steel capital of North America. Afterwards it fed upon the expanding cities of the Middle West, and they upon it. It therefore looked in two directions. Principally owing to the Pittsburgh complex, America was able to proclaim at the end of the nineteenth century that it led the world not merely in steel production but also in coal output. In the decades which followed, other steel cities —Buffalo, Toledo, Cleveland, Chicago and Birmingham—were to challenge Pittsburgh.

America has abundant and widely distributed sources of energy; but the concentration of industry rooted in the north-east a century ago was to show no significant shifts in response to the opening of new mineral deposits. Petroleum, first exploited commercially in 1859, was also available in Pennsylvania. Its use was limited until the development of the internal-combustion engine. Thereafter, the search for new supplies extended, wild-cat speculation followed and the older producing states of the north-east yielded to Louisiana, Oklahoma and Texas. As the railways had manifold effects upon coal mining, so the automobile had its consequences for oil production. Oil was eventually to compete with coal as an industrial fuel and as a source of energy for transport. While coal called for elaborate rail or water carriage, oil fuel could flow. The shift of the principal centre of production was therefore less significant. Though the first was laid in 1865, pipe-lines belong to a later phase of America's industrial revolution. So, too, does the large-scale commercial exploitation of natural gas. Pipe-lines for natural gas distribution were eventually to outnumber those for coal gas. Coal, petroleum and natural gas more than held their own in the provision of thermo-electricity in the manufacturing north-east. The phase of hydro-electric power was to have no great importance for it.

Industrial change also affected the processing of farm products. Canning increased rapidly after the Civil War and refrigeration was introduced into meat marketing in the 1880's. Giant industries such as those of Swift and Armour were established in the fattening lands of the Middle West. The agriculture of the southern states also became aware of the pulse of industry. In addition to the establishment of the steel industry in the Birmingham area, the processing of staple crops was initiated. Backwardness, which had been the concern of Daniel Goodhoe's *Enquiry into the causes which retard the Southern States* (1848), was persistent. However, workshop and factory took over tobacco manufacture; while cotton mills began to migrate from New England to the cotton country. There might be marked regional variations in the speed of evolution of industry, but by the end of the century, with industrial output exceeding that of farm output in value, the U.S.A. had achieved a new status as a manufacturing nation. In 1894 it was able to claim that it led the world in the absolute volume of its industrial output, and leading the world meant that its production amounted to half as much as that of all Europe. In the final instance, demand promoted this. The American market, especially suited to the large-scale production of standardized commodities, fostered and furthered the development of industry. The years which ushered in this new status were called 'The Gilded Age'. The black smoke of America's industry yielded more gold than the white road to the Yukon which beckoned in the same decade.

The cities which arose to accommodate industry evolved certain common features. First, there was a superficial uniformity in which was balanced a curious order and disorder. Order was rooted in the grid-iron street plan: disorder sprang from the traditional lack of controls within it. Identity was imparted by the tree-lined avenues which belong to the cities of the humid and semi-arid lands alike. Industrial Scranton, viewed from its surrounding mountains, gives the illusion of a spa; while the cities of the High Plains are oases of green. The planting of avenues, especially of elm, also took place sufficiently long ago for the twentieth century to reap the benefit of their maturity (and responsibility for their replacement). Extensiveness is a second feature of the American city: the area covered is almost always greater per head of population than in Europe. Even Sinclair Lewis's Zenith in *Babbitt* was 'built—it seemed—for giants'. A third feature is the skyscraper core—explained more by prestige than by high land values in the Central Business District. New York led the way with skyscrapers in the 1890's. The shanty-town fringe, seeking out cheap land beyond the limits of the urban ratable area, is often its counterpart.

The garden city and satellite city have developed in America as in Europe; while, in an attempt to relieve the traffic problems of the city core, the establishment of extra-urban shopping centres goes on apace. Fourthly, by comparison with the European city, the American is cosmopolitan. Economic differentiation is accompanied by ethnographic. Ethnographic differentiation is commonly tied to the geometric pattern of blocks, and changes in its pattern are one of the principal problems of North America's urban areas.

Beyond these common features, setting and siting convey a measure of uniqueness to the individual town, down to the detail of its principal building materials. The rich diversity of stones available in the hinterland of the 'tidewater cities' finds expression in their façades. Washington, for example, is a city with an Athenian or Roman prodigality of marble; Baltimore is typified by its 'brownstone' (sandstone) houses. The abundance of timber in the eastern half of the U.S.A. is reflected in the large number of 'frame houses', and the persistent wooden buildings in North American urban places imply correspondingly high fire risks. The sod house of the timberless plains has disappeared, but the split-log cabin of the eastern pioneer still lingers and the adobe dwelling of such a city as Santa Fé commands a scarcity value. History, relatively shallow though its roots may be, still counts for something in the appearance of American cities, as Constance Green has illustrated in her selected studies.

Yet speed of growth and speed of change make a greater impact. Few countries in the world have mechanized their building industry as much as the U.S.A.; few have shown such a sensitive concern about 'real estate'. The ingenuity with which problems of terrain and climate are overcome is matched by the flexibility in constructional methods which are a response to this speed. The continent thrills at its boom towns, which are usually explained by the discovery of new resources (e.g. Edmonton and oil), but which may also be the result of changing fashion (e.g. Floridan resorts, with the rise of winter holidays after 1900). Ghost towns are the complements of boom towns, though geographers have yet to invest them with the academic respectability given to deserted medieval villages in England. While ghost towns emphasize one aspect of the speed of change, another is found in the redevelopment of city centres. Redevelopment is one means of assessing urban vigour. The post-war transformation of down-town Pittsburgh provides an example. Speed of growth and speed of change offer challenges to architects, and there is abundant evidence to support the claim that the architecture of the U.S.A. is the most progressive in the world. Academic investigation of urban areas has also given rise to a classical literature, and no one has been more imaginative in his approach to urban studies than Lewis Mumford. The rise and fall of civilizations is measured in the growth and regeneration of a country's cities.

THE EMERGENCE OF CANADA

In a nutshell Canada consists of four natural settlement areas separated by three major barriers. From east to west the settlement areas are the Atlantic coastlands, the St. Lawrence Valley with the Great Lakes peninsula, the Prairies and the Pacific coastlands. These are interrupted by the 400-mile barrier of the Appalachians, the 900 miles of inhospitable Shield and by the 400 miles of the western Cordillera. To the south there is continuity of settlement with the U.S.A., though the political boundary interposes a very real obstacle. To the north, settlement thins out rapidly in a vast frontier zone. The settlers who have occupied the land fall into five principal groups. Reviewed in the light of contemporary census returns, they consist of two markedly non-Anglo-Saxon groups sandwiched between three dominantly Anglo-Saxon groups. The geographer's concern is the interrelationship between the physical design and the human distribution. What form has it taken?

In the days when Canada was called a Dominion, Lilian Knowles wrote:

> The Dominion of Canada started from six different points. Her growth has not been brought about by pushing back the land frontier as in the U.S.A.: it has been a process of coalescence of different starting points by the filling in of intermediate spaces.

The six points were defined as the Atlantic coast, the Quebec lowlands, the Hudson Bay lowlands, the British Columbian coast, Upper Canada and the Red River Valley of Manitoba. Lilian Knowles looked at them as an historian, but the geographer would emphasize that the first four are coastal and the last two continental. The first stage in Canadian evolution is the rise of the coastal centres; the second, the establishment of the inland centres; the third, the shift in importance from the seaboard to the interior centres. It is proposed to review briefly the six points of settlement identified by Lilian Knowles and to comment on the process of fusion between them.

The name Canada first appeared on the Harleian portolan of 1536 and

V DRUMLIN FIELD IN SOUTH-CENTRAL WISCONSIN

The graining of the drumlinoid features from the western edge of Dodge
County in Wisconsin strongly influences field boundaries and farming operations. The original woodland with interspersed prairies has been transformed
into crop and pasture country

VI MISSISSIPPI LOWLAND, LOUISIANA

The pattern of long lot divisions stretches back from the waterfront of Bayou Lafourche in this area of French-speaking settlement from Assumption parish (La.). The river is lifted above the countryside by its levées. Sugar cane, corn and rice are the chief crops on its fertile delta soils

was written over the St. Lawrence lands. The curtain rises with the 1534 reconnaissance of the St. Malo pilot, Jacques Cartier, and the scene for the establishment of New France was broadened through Champlain. Quebec, Three Rivers and Montreal were the first centres of a settlement pattern which clung tenuously to the river line. The church and army buttressed a farming community drawn principally from Normandy, Brittany, Biscay and the Île de France. Trapping and casual trading by the *coureurs de bois* engaged as many as a fifth of those who were based on the St. Lawrence shorelands. Canada was born on the banks of the St. Lawrence; its French-speaking population was small when Wolfe overcame Quebec, but it refused absorption.

The British foothold on the Atlantic seaboard lay in peninsular Nova Scotia, which supported fortress outposts against those of New France. The removal of the military threat and the eviction of French Canada's Acadians from the Fundy lands led to the strengthening of British settlement. Mainland New Brunswick and offshore Prince Edward Island completed the trio of provinces which drew wealth from forests and fisheries to supplement their farming. A strong Scottish element, partly reflecting late-eighteenth-century depopulation of the Highlands and Islands, was joined by immigrant Loyalists. The three maritime provinces still had nearly as many inhabitants as Quebec and more than Ontario down to the census of 1850.

Canada's third settlement was Hudson Bay. In 1670 there was founded in London by Prince Rupert and his Seventeen Associates a 'Company of Adventurers of England trading into Hudson's Bay'. The names around the approaches to this great icy embayment into which much of Canada drains, tell the saga of fruitless, often tragic journeys in search of a north-west passage. The Hudson's Bay Company, however, enjoyed success, sailing safely on its summer journeys through northern seas to the wooden fortresses of Prince Rupert's land where the harvest of furs was assembled. The Hudson Bay and its tributary lands, which became a virtual empire of the company, never supported permanent settlement of consequence, but they played a significant role in the evolution of the realm.

The fourth seaboard group of settlements lay a continent away on the Pacific. The Hudson's Bay Company had extended its tentacles to western shores, but permanent white settlement was tardy in establishing itself on Vancouver Island and around the Fraser estuary. Even after a boundary was agreed with the U.S.A. the threat of absorption remained. British Columbia and Vancouver Island became crown colonies in 1859, with New Westminster and Victoria as their respective administration centres.

I

They were reunited in 1866 and given provincial status in 1871, but the way home or to eastern Canada was still around Cape Horn.

The interior settlements grew slowly. Those of Upper Canada showed more active development after the American revolution, when Loyalist groups concentrated about the strong points of Kingston, Fort York and Niagara, to become the force behind the anglicization of Canada. The close of the eighteenth century also witnessed the subjugation of the Ontarian Indians and their retreat from the lakeshore lands. A formal land survey was initiated and the soldiery was provided with farmsteads (200 acres for a private soldier; up to 5,000 acres for an officer). With the return of peace in Europe immigration increased, later to be spurred by Gibbon Wakefield and the Colonization Society.

The year 1825 saw the foundation of The Canada Company:

for the purpose of purchasing, settling & disposing of waste & other lands & for making advances of capital to settlers on such lands, for the opening, making, improving, & maintaining of roads, & other internal communications for the benefit thereof.

But Upper Canada was not easy of access, and it still took a week or more by boat and portage from Montreal to Kingston in 1810. If geography caused detachment, ethnography exaggerated it. Lower Canada, finely portrayed by Joseph Bouchette in his *Topographical Description of Lower Canada* (1815), had little in common with Upper Canada. The *Durham Report* of 1835 was expressive of Canada's growing pains. Upper and Lower Canada might be joined in an uneasy political union in 1841, steamboats might strive to link lakeside settlements, canals might be cut, but the coalescence of Canada's several settlements made slow progress.

The second of the continental sites was in the Red River Valley. On the edge of the great sea of grass, in 1811, Lord Selkirk purchased 116,000 square miles from the Hudson's Bay Company, established a joint stock company and issued a *Prospectus of the New Colony*. The initial entry was by way of Hudson Bay and the Nelson River, which took a year and meant over-wintering at one of the Bay settlements. The colony was an adventurous undertaking and Rhodesian in concept. But, in its early stages, it was a failure. Prairie fires, climatic hazards, wolves, coyotes, grasshoppers, half-breeds and the rivalries of fur companies left the first settlers bereft of the means of living. The problems of access were manifold and there was little real contact with the peopled parts of Canada.

The definition of Canada (Figure 19) has been part and parcel of the

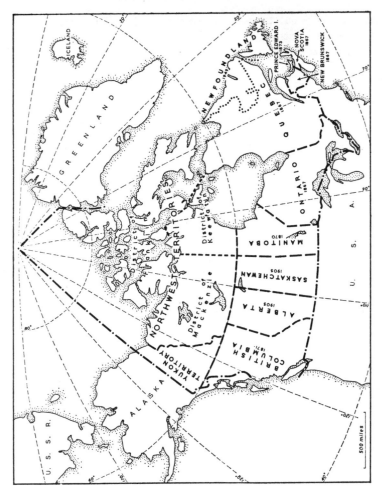

Figure 19 *Definition of
the Provinces of Canada*

The emergence of the pro-
vinces of Canada has been a
more simple process than the
emergence of the states south
of the border. The North-
lands are now formally divided
into districts and territories

definition of the northern boundary of the U.S.A. The earlier evolution of Canada's settlements, especially those of the interior, was closely related to that of lands south of the border. Moreover, evolution has been inseparable from likenesses and differences in the physical landscape north and south of the border.

The most important physical difference is that Canada lacks a counter-part to America's Middle West. The place of America's fertile heartland with its natural ease of movement is taken by the lands of the Shield. Southern Ontario is in some respects a physical prolongation of the Middle West; the southern half of the Red River Valley is in Minnesota and North Dakota. The occupation and development of the Middle West automatically affected these two areas. First of all, the barrier of the Shield diverted the stream of immigration from the north-west to the south-west, so that many who came to settle in Canada moved into the U.S.A. and the offspring of Canadian pioneers followed suit. Secondly, deflection of migrants from the north-west meant that the so-called 'Prairies' re-ceived their first flood of colonists from the south.

Early railway development only served to exaggerate the stresses and strains of allegiance. The first Canadian lines, from Toronto to Montreal (1856), were extended to Portland (Maine) and Chicago (1859). The first lines into the Red River Valley settlements ran from Minneapolis/St. Paul to Winnipeg. The 'All Canada route', by way of the Great Lakes to Fort William and by lake and portage to the Manitoban gateway, was a poor alternative.

By the British North America Act (1867), Canada achieved federa-tion and political independence as the first Dominion. The following year a trans-Canada railroad was begun, which was to take 15 years to complete. The iron horse, the iron rail, the iron bridge and the capital for them had to be imported, for Canada had as yet no adequate iron and steel industry. The saga of the railway's assault on the Laurentia, its march across the Indian-infested plains and its break through the three-fold ranges of the western Cordillera is one of the great engineering epics of the nineteenth century. Canada proved to be very much what H. G. Wells called 'a railway country'. When Vancouver was joined by the line of steel the 15-year-old province of British Columbia became an effective part of the federation.

Railway construction taught the Canadians much about Canada. Engineering called for the application of scientific knowledge and this in turn led to a fuller understanding of the land. The surveyors who planned the railway routes also revealed more fully the true nature of Canada's climates. In adapting their work to new experiences of heat and cold, of downpour and blizzard, the railroad planner and operator had tasks before them unknown to those in the kindlier climes of Atlantic Europe.

In retrospect, it may be said that most of Canada has been peopled by way of its rail routes. Alternative methods of transport have been added; but the railroad remains a fundamental piece of apparatus in Canadian commercial life. It has helped to overcome the natural barriers within Canada, to offset human response to the north-south grain of the country, and to strike a closer bond between the essentially 'English' settlements of the Maritimes, Ontario and British Columbia. This bond was the more important because the railroad provided the means of entry for the emigrants who peopled Manitoba (admitted as a province in 1871) and broad Assiniboia, which was later divided into the provinces of Saskatchewan and Alberta (1905). Only a third of the settlers which it carried prairiewards during the peak years of recruitment were of British stock. The rest came principally from the U.S.A. and from central and eastern Europe.

American migration to the Prairies was important for several reasons. First, it was undertaken by immigrants who were used to the broad horizons of the New World and whose reaction to the open prairies was likely to be positive. Immigrant Europeans often had psychological as well as physical problems of adjustment, unless they were accustomed to the open landscapes of the eastern steppes or were of closed religious communities. The frontiers of settlement demanded adventurers and speculators, while semi-arid lands called for experience with irrigation and dry farming (such as the Mormons who came into south-west Alberta from Utah had already acquired). Secondly, the American invasion might raise problems of allegiance at the national level, but it created a great web of affiliations at the personal level. In the mid-nineteenth century many Middle Western Americans were of Canadian origin; fifty years later Prairie Canadians had parallel kinship in the Middle West. The process was a part of the 'greatest reciprocity movement in history'.

For Canada, the westward course of empire was to be succeeded by a northward course. The first advance into the northland was sustained, if not promoted, by the railway. The Canadian Pacific Railway was a successful commercial venture, but the complementary Canadian Northern Railway, which struck an east-west base line through higher latitudes, was taken over by the federal government as the Canadian National Railway in 1916. It was fundamental to the settlement of the Clay Belt of Ontario and Quebec, to the peopling of the northern margins of the Prairie

Provinces (latitude 60°N was defined as their boundary when the North-West Territories were given provincial status in 1912) and to the occupation of northern British Columbia. The Yukon has no rail link.

The occupation of the greater part of Canada has taken place in relatively recent times. Records which illuminate the detail and process of settlement and survey are comparatively abundant. Physical records embrace those of the Canadian Geological Survey (established 1843), which published its first great volume in 1863. Human records are still being garnered by the Champlain Society. Their range is wide—from Acadian descriptions by the Sieur de Dièreville to Sagard's missionary journey to the Huron country, from the letters of early Ontarian loyalists to the Hargrave records on the old North-West, from Simpson's western journals to the Port Vancouver letters of McLoughlin from 1844 to 1846. In such records repose the seeds of nationality. Canada has travelled a long way from their records to its newly computerised Inventory of land use and land capability.

Mechanical unity may have been one means which has enabled Canada to overcome separatist tendencies, but the unity of thought which derives from sharing common experiences is equally important. Canada's dispersed and disparate tracts of primary settlement have passed beyond the stage of coalescence. The task today is to strengthen community of feeling in the narrow zone of settlement sandwiched between the northern frontiers of a natural world and the southern boundary of an economic giant. For the modern Canadian it is all too easy to shed responsibility for the control of this national unit, by escaping into the adventures of the northland or deserting in favour of greater material gain to lands south of the border. Nationality is not easily created, but its seeds are found in records from the past as well as in triumphs of the present.

THE RECORD OF THE PEOPLE

The most important guide to population in the modern state is the census: according to its preliminary estimates for 1960–1, the U.S.A. has 179,000,000 inhabitants and Canada 18,000,000. The great bulk of the population is concentrated in the north-east (Figure 20). In the U.S.A. the two most populous administrative units are New York State and California: in Canada, Ontario and Quebec. Censuses have been taken in the U.S.A. since 1790, while Canada can lay claim to some of the oldest census records in the world.

The first Canadian population census, initiated by the Intendant Talon in 1665–6, resulted in the dispatch of a 154-page manuscript to France. It was followed immediately by a farm survey, in which were recorded the number of *arpentes* under cultivation, grain yields and stock numbers. Authorities anxious about population numbers, their resource base and technical equipment, conducted more than thirty censuses before the

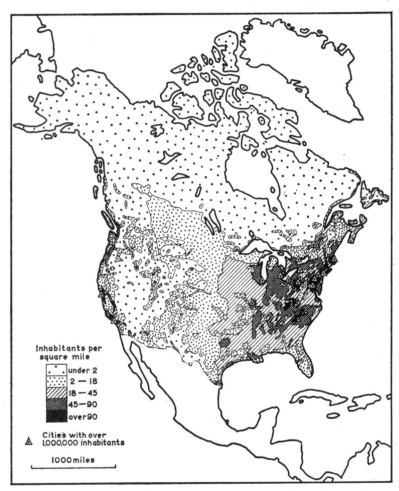

Figure 20 *Population Density in the U.S.A. and Canada*

The concentration of population in the east centre and the extensiveness of the thinly inhabited tracts are the two predominant characteristics of North American population distribution (based in part on Goode's *World Atlas*, Chicago, 1953)

French retired from Quebec. The first 90 years of British rule yielded a meagre statistical record, but in 1851 an act was passed providing for a decennial census. The preambles to the ensuing publications are often as readable as their statistical contents are revealing. Reference may be made to two of them.

The introduction to the 1851 census suggests that Canadian authorities looked intently south of the border to see what lessons might be drawn from a comparison with American returns. Ohio set the standard and, for the betterment of Canadian farming, it was urged that new milk-cow breeds should be bought, more sheep should be kept, more cheese should be made, more clover, grass, Indian corn and tobacco should be grown. American migration patterns were also observed, perhaps forecasting those of Canada.

> Internal emigration moves upon the same parallel of latitude, and this fixed law is found to be almost universal, and it may probably arise from the fact that men naturally seek, as much as possible, to preserve the same climate, habits and social institutions to which they have been accustomed.

The 1871 volumes of the Canadian census are a virtual Old Testament, in that they include lists of all the early censuses and were the last to exclude the Western provinces. Their compilers acknowledged them to be 'the sole contribution which our country can offer to the science of human life'. The significance of industry for Canada was recognized by the addition of an industrial census in 1910. Today, facts and figures are assembled for forecasting the future as well as understanding the past. For any country which sells farm products as futures, situation reports are vital. Experimental officers, farmers, managers of elevators, postmasters and many others are enrolled in their thousands to compile monthly reports. The assembly of statistics has become a major national undertaking and the Census Office of Canada issues 500-page volumes today for every ten pages of Talon's report of 300 years ago. At the same time, a Canada Land Inventory, described as a computer-based land data bank, has been established.

The first federal census of the U.S.A. was a simple numerical statement which listed males and females, adults and minors (under sixteen), slaves and free. The third census, of 1810, introduced 'an attempt to show the number, nature, extent, value and situation of the arts and manufactures'. Information was not provided for the detail of county and township until after 1830, and employment statistics were delayed until a decade later. From 1860 onwards, country of birth was identified for foreign-born population and preambles became of increasing interest. The six volumes of 1850 had become twenty-five volumes by 1891; today there are several volumes for almost every state. Successive censuses enable the broad evolution of the U.S.A. to be traced, although changes in the methods of collection from decade to decade restrict comparability (Figure 21).

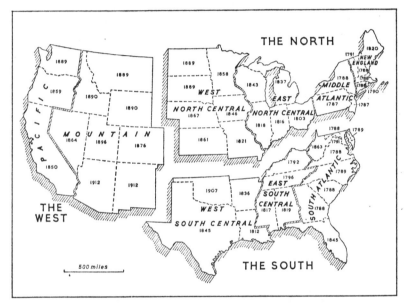

Figure 21 *Census Units of the U.S.A.*
These are the divisions into which the U.S.A. has been separated cartographically for the purpose of regional statements about the census. The map also shows the dates of admission to the union of individual states

. Contemporary preambles are models of conciseness and include definitions of primary importance for the human geographer concerned with the U.S.A. Brief reference to several examples from the population volume must suffice here. The concept of 'race', which the Bureau of the Census applies rather to nationality than to 'biological stock', provides the first example. Population is formally divided into white and non-white, and the term Negro includes 'persons of mixed white and Negro parentage'. Mixtures may be very involved and those of 'non-white races are classified according to the race of the father'. The classification of population according to employment status provides a second example of the care given to definition. The American labour force (unlike that of certain European countries) does not include 'persons primarily occupied in their own housework'. The occupational classification system used in 1950 contained 469 items, grouped under twelve major headings. The detailed items included categories of occupations strange to English ears (but no more strange than many items confronting an American in the Ministry of Labour's *Classified Index of Occupations*). There is a breath of the natural world about occupations such as chainmen, rodmen, axmen, raftsmen, woodchoppers, sawyers, powdermen and blasters. The differentiation of urban and rural settlement, an especially debatable matter in the New World, provides a third example. For the purposes of the 1950 census, urban population was defined as comprising all persons living in

(a) places of 2,500 inhabitants or more incorporated as cities, boroughs, and villages, (b) incorporated towns of 2,500 inhabitants or more except in New England, New York, and Wisconsin, where 'towns' are simply minor civil divisions of counties, (c) the densely settled urban fringe, including both incorporated and unincorporated areas, around cities of 50,000 or more, and (d) unincorporated places of 2,500 inhabitants or more outside any urban fringe.

The concept of the Standard Metropolitan Area was also introduced to assist in social and economic survey. It was defined as 'the entire population in and around a city of 50,000 inhabitants or more whose activities form an integrated social and economic system'.

These examples of definition from a single census volume indicate the tendency towards precision in expression which characterizes the records of the North American people. The censuses of farming and of manufacture are the great counterparts which complement those of the people and their characteristics. These volumes are what Adam Smith would have called a true statement of the 'wealth of nations' and they are the raw material to which the economic geographer turns.

.   .   .   .

European man has occupied and transformed the sub-continent of North America in not much more than four centuries. In the process he has himself been transformed into something other than a European. Out of the amalgam of the old nations which set foot in North America have been created new. Among them, the U.S.A. is so predominant that Canada (although it is larger) and Mexico (despite its longer history) play subordinate roles. The peopling of the sub-continent is not complete. The U.S.A. has empty spaces and Canada has more unoccupied than occupied land. Advance into new territory and retreat from lands of long occupation proceed simultaneously, for both types of country are revalued with changing techniques. These facts are the more important because North American man is particularly swift to adopt and to adapt new practices.

The record of the land has been reviewed in Chapter 2: the record of the people concludes this chapter. Both records help towards understanding the causes of many of North America's contemporary problems. Fuller measurement and assessment of man and the land are the natural objectives of a sub-continent that is more concerned with forecasting the future than probing the past.

SELECTED READING

There is virtually no limit to the additional reading which springs out of this section. Two indispensable texts are

C. O. Paulin, *Atlas of the Historical Geography of the U.S.A.* (New York, 1932)
R. H. Brown, *Historical Geography of the U.S.A.* (New York, 1948)

Some idea of the richness of United States map material can be gleaned from

C. le Gear, *United States Atlases*, Vol. I (Washington, 1950), Vol II (Washington 1953)

Useful reference for past facts and figures are

Historical Statistics of the U.S., 1789–1945 (Washington, 1949)
Historical Statistics of Canada (Cambridge, 1964)

the dry bones of which are given life by such studies as

E. C. Kirkland, *A History of American Economic Life* (New York, 1941)
Frank Thistlethwaite, *The Great Experiment* (Cambridge, 1955)
J. Wreford Watson, 'Canadian Regionalism in Life and Letters', *G.J.*, 1965, 131, 21–33

Glimpses of the prehistoric scene are provided in
>   C. O. Sauer, 'A Geographical Sketch of Early Man in America', *G.R.* 34 (1944), 529–73 and 'The Early Relation of Man to Plants and Animals', *G.R.* 37 (1947), 1–25
>   T. Heyerdahl, 'Plant evidence for contacts with North America before Columbus', *Antiquity* 38 (1964), 120–33

and the native scene should be compared in
>   C. Wissler, *Indians of the United States* (New York, 1940)
>   A. L. Kroeber, *Cultural and Natural Areas of Native North America* (Berkeley, 1947)

The story of Federally owned land in U.S.A. is traced in
>   R. M. Robbins *Our Landed Heritage, the Public Domain, 1776–1936* (New York, 1942)

while details of the administration of imperial land regulations for Canada appear in
>   *Canada, 1763–1841, Immigration and Settlement* (London, 1939)
>   L. F. Gates, *Land Policies in Upper Canada* (Toronto, 1968)
>   Alan Wilson, *The Clergy Reserves of Upper Canada* (Toronto, 1967)

Boundary definition is covered in
>   N. L. Nicholson, *The Boundaries of Canada, its Provinces and Territories* (Ottawa, 1954)

For sample urban vignettes of the U.S.A., see
>   Constance M. Green, *American Cities in the Growth of the Nation* (London, 1957)
>   Raymond E. Murphy, *The American City, an urban Geography*, New York, 1966)
>   John W. Reps, *The Making of Urban America*, (Princeton, 1965)

A unique construction for 1810 is offered in
>   T. P. Keystone (alias R. H. Brown), *Mirror for Americans: Likeness of the Eastern Seaboard* (New York, 1943)

Three books which illustrate a contrasting approach to the landscape are
>   F. Harper, *The Travels of William Bartram* (New Haven, 1958)
>   H. R. Merrens, *Colonial North Carolina in the Eighteenth Century* (Chapel Hill, 1964)
>   L. de Vorsey, *The Indian Boundary Line in the Southern Colonies, 1763–75* (Chapel Hill, 1966)

A classical study of the Eskimos and their habitat is
>   J. Louis Giddings, *Ancient Men of the Arctic* (London, 1968)

And a final piece of bedside reading
>   Van Wyck Brooks, *The World of Washington Irving* (New York, 1964)

The United States

4

The Industrial North-East and the Problem
of Maintenance

*The North-East Quadrant—The Tidewater Lands—New England—
New York: the City and its Tributary area—The Power of the
Hinterland—'Je maintiendrai'.*

THE NORTH-EAST QUADRANT

Appalachia, one of the grand natural regions of North America, is in
detail one of the most complex. Its complexities are nowhere greater than
in the north-east quadrant of the U.S.A., where the country is invested
with its largest concentration of population and wealth. To the east,
Appalachia advances into the sea through the New England states; to the
west, in Ohio, the dissected Allegheny plateau gradually yields to the
quieter landscapes of the Middle West. Centrally situated are New York
State and Pennsylvania ('The Keystone State') which span the isthmian
area between the Atlantic and the lower Great Lakes. Between the fiorded
entrance to Hudson River and the multiple ria of Chesapeake Bay an
incipient coastal plain is born and shared administratively by New Jersey,
Delaware and Maryland. Delaware is so low-lying that it is facetiously
described as consisting of three counties at low water and two at high
tide. Virginia is the transitional state to the south, with mountainous
West Virginia on its margin. Diversity of landform, variety of resource,
age of settlement, accumulation of capital and experience, proximity to
the most progressive part of the Old World—all made their contribution
to the early leadership of the north-east. As in Old World Hercynia,
topography has exaggerated contrasts in the occupation and development
of the land, so that sleepy hollows, hill-billy valleys and outlaws' eyries
are not so far removed from the hearths of North American manufacturing, the mercantile metropolis of New York and the political capital of
Washington. In the commercial foreland, from Washington (D.C.) to

Boston (Mass.), there has evolved one of the world's most remarkable urban phenomena, a virtually continuous built-up area. This was christened in a book by Jean Gottman published in 1961 and bearing the title *Megalopolis*. The characteristics of Megalopolis spring from population concentration, its associated man-made landscape, and elaborate systems of circulation. Today, its population approaches 40 millions (37 millions in 1960)—roughly the equivalent of that of the Argentine and about twice that of Canada. The continuous urban and suburban area of Megalopolis (cf. Figure 23) was the first of its kind in the world and is still the largest. It has generated and continues to generate great wealth, with correspondingly powerful national and international controls. As the first phenomenon of its kind it suffers the first problems of its kind.

Megalopolis, though polynucleal and though divided administratively and set against a varied physical terrain, has a human unity. At the outset the interplay of land and water caused problems. Inlets have had to be circumvented, slow-moving ferries to be tolerated, sudden floods in the shallow river beds to be contended. Long and expensive bridges could only be constructed with the improvement of the suspension principle. The 'Pulaski Skyway', which leapfrogs the Hackensack marshes in New Jersey, the George Washington Bridge over the Hudson and the Wilmington Bridge over the Delaware reflect the toll of interrupting waterways. Tunnels are the alternative. Manhattan Island claims most, but underground roadways have also enabled Baltimore to escape the obstruction of the Patapsco estuary. A primary fact in the history of the north-east quadrant has been that the rivers of the Atlantic slope have not been readily navigable. Only the Hudson below Cohoes Falls has been of value for more than canoe and portage. It is the valleyways rather than the waterways that have been the integrating features.

The most remarkable route is that opened by the transverse Hudson, extended northwards by way of lakes George and Champlain to the St. Lawrence Valley, and with its Mohawk tributary forcing a gap between the Adirondacks and Catskills at 500 feet a.s.l. to the Lake Ontario coastlands. Back from the Middle Atlantic coastal plain, the Delaware, Susquehanna and Potomac with their tributaries offer negotiable if tortuous ways to the headstreams of the Ohio river system. But the Erie railroad has to climb to 1,000 feet near Binghampton N.Y. and the Baltimore and Ohio railroad to well over 2,000 feet to negotiate the central Appalachians. Farther south, in Virginia, the James River opens routeways into West Virginia via the New Greenbrier rivers. New England stands aside from this system of linking routes. Modern road-

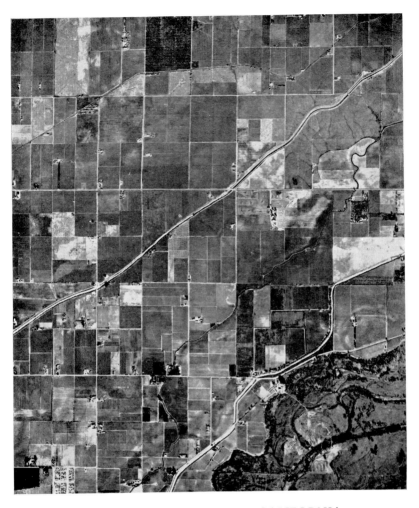

VII SAN JOAQUIN VALLEY, CALIFORNIA

The flat floor of the Central Valley of California in Fresno County lends itself to evenly spaced holdings. Natural waterways, underground water and, latterly, major irrigation canals, provide for fruit, cotton and truck farming on small farm properties. The contrast with the unirrigated slopes in the south-east corner is vivid

VIII THE WASATCH FRONT IN UTAH

The juxtaposition of semi-arid, thinly wooded or meagrely grassed mountain ribs and irrigated basin land characterizes the faulted front of the Wasatch range in the vicinity of Logan (Utah)

building techniques have reduced dependence upon natural conditions. The most recent arterial highways, such as the Pennsylvania Turnpike, pursue virtually Roman-straight courses, tunnelling as well as viaducting.

In general, the form of the land has not favoured the employment of canals. To facilitate waterborne traffic, two isthmuses have been bisected— one by the Chesapeake-Delaware Canal, the other by the Cape Cod Canal. Both canals can accommodate ocean-going shipping. The New York State Barge Canal (formerly the Erie Canal) has played a much more historic role in unifying the area. Heavy-load traffic moves over it during the 8 months of open water. Beyond the range, the deeply entrenched Ohio and its tributaries have been canalized for of fuel.

In this North-East Quadrant several constituents may be identified. They differ in their local resource base and in the detail of their evolution. The six New England states form a distinctive unit, to be balanced by the less well-defined 'Tidewater Lands'. At the meeting ground of the two is the metropolitan area of New York with its historic route to the west. The hinterland includes the established centres of heavy industry in the Upper Ohio Valley with their closely related Lake Erie outliers. This hinterland finds the Atlantic seaboard to the Middle West and it is here that the so-called Manufacturing Belt finds its fullest expression.

THE TIDEWATER LANDS

'The maritime zone of every continent enjoys a superiority over all others,' wrote the Swiss immigrant Arnold Guyot at Boston in 1849. 'It is in this region of contrast between the land and the ocean that life is unfolded in its most intense and diversified forms.' His opinion is the more valid for the Tidewater Lands because they are backed by a hinterland in which the variety of resource is rivalled only by that of California. Diversity of rockform and variety of vegetation have been matched by the historic mingling of native husbandry, British and European stocks, crops and methods in agriculture and by the fullest expression of industrial ingenuity in the field of manufacture. Moreover, the republic was born here and the centre of administration resides here.

The Tidewater Lands embrace a much-embayed foreland, with sand-bars and spits ponding back lagoons on the Atlantic seaboard, with shallow sand and loam-surfaced terraces divided between pine barrens and market gardening, lifted above marshlands rich in bird life and tidal flats with a wealth of shellfish. Formerly, the piney lands were neglected, but

K

today they are the scene of lively forest improvement, active cash cropping, coastal reclamation and thriving recreational areas. Peninsularity lengthens and strengthens growing conditions. South of the Patapsco estuary, as Volney observed, farmsteads no longer sport a sleigh. Proximity of the foreland to the big cities has favoured the rise of business men's estates which make George Washington's Mount Vernon look decidedly modest.

At the head of navigation are the seaboard cities, commonly at sites which mark the appearance of the 'fall zone'. Falls in Philadelphia and Gwynn Falls in Baltimore have today been incorporated in city parks. Back from the 'fall zone' occur a series of basins of sedimentary rocks, with fertile and generally well-drained soils. Limestones and sandstones are quarried, cement plants process the calcareous rocks, glass works (some of the largest in America) exploit local sands, brick kilns employ local clays, porcelain factories use local kaolin (though its quality is not high) and chemical plants consume local salt. The variety of flora which so greatly excited Carl Linnaeus's pupils, contributed richly to early American *materia medica* and gave birth to a vigorous school of naturalists around John and William Bartram in Philadelphia, pointed to the multiplicity of opportunities for cultivated plants. Here, indeed, some of America's finest cultivable land was occupied by some of the New World's best farmers (Plate III). William Penn's woods have been partly transformed into tens of thousands of family-farms of about 80 acres in size, where grain and fodder crops support dairying and poultry raising, where apples and pears are mixed with peaches, tobacco, vines and mulberries. In this sylvan country the line of the forest is on the edge of the ploughed land as well as on the blue-ridged horizon. On summer nights fireflies rise in clouds to its tree-tops before they disappear.

Back from the Tidewater Lands, Appalachia becomes more classically furrowed than in any other section. The industrial absorption of farmland along the tidewater edge is matched by a retreat of the frontier of agriculture in the hinterland. Yet there are fertile, well-stocked limestone grasslands, as in the Shenandoah Valley; while timbers (oak, beech, elm, basswood, tulipwood, sassafras and half a dozen others) are much esteemed for constructional purposes and for furniture making. Rocks are selectively quarried—granites, marbles, limestones and slates—so that Washington displays a Mediterranean profusion of marble facings in its monumental buildings, which contrasts with the humbler 'brownstone' of Baltimore. The wild life of inner Appalachia, from soft-shelled and snapping turtles, through spade-footed toads and five-lined skunks, to 'protected' brown bears, pads through mountains with magic names. Indeed, few parts of America are spiced with such an appealing array of place-names.

For all its agricultural opportunities and rural retreats, the census proclaims this to be a country of urban dwellers and factory employees. The seaboard cities reach their maximum size in Philadelphia, Baltimore and Washington. Philadelphia, 'the city of brotherly love' conceived and planned by William Penn, has an orderly but cramped eighteenth-century core to which have been added a multiplicity of late nineteenth-century workshops, impressive cultural monuments and extended industrial tracts on the reclaimed waterfront, with outliers in Wilmington and Chester. Its industry and trade are soundly based, though there is a substantial import of raw materials—oil, sugar, jute, fibres, ores, pulp, fuels, china clay. It has all 57 varieties of the foodstuff industries and more; while at hand, on the Delaware estuary, is the Fairless iron and steel plant. A long-established tradition lends prestige to some pursuits—printing and publishing, glass and china manufacture, the clothing industry, banking and brewing. Philadelphia remains one of America's most prosperous cities.

Baltimore, an hour away by multi-track railroad or turnpike highway, is a heavyweight in commerce. It retains historic milling and grain exporting interests, has immense oil imports from the Gulf coast, raw sugar from Cuba and the Philippines, iron ore from a dozen countries, pyrites, pulp and timber. It is a leading coal exporter. The 'trade winds' that blow across the city on a hot summer night are part of the New World's olfactory geography. At Sparrow's Point, Baltimore claims to have the world's largest tidewater steel mill. Tin plate, ship's plate, copper and cables are leading products: superphosphates and acids, valued by-products. Shipping interests are manifold and, depending upon the chosen criteria of assessment, it is at least America's third port.

By comparison with these cities of producers, Washington is very much a city of consumers. The planned capital, 'a city of magnificent distances' to Charles Dickens, is a fitting centre from which to control the magnificent destiny of the nation—at least when the Japanese cherries bloom in spring or the grapes hang on Georgetown villas in autumn, for its summers are sub-tropical and its winters sub-human. Artificial climatic control through air-conditioning has transformed its working conditions. The scale of America's mid-twentieth-century responsibilities has forced on the city's expansion and the very word 'Pentagon' has become a synonym for a mammoth administrative block.

The constellation of cities is extended north-eastwards to Trenton, Elizabeth and Newark in New Jersey, while the logical southern continuation is in the 'fall zone' cities of Richmond (Virginia's state capital), Petersburg and Frederiksburg. A well-distributed group of moderately sized cities, former market towns which have acquired manufacturing industry, occurs inland. Reading is a textile town; Allentown, a cement city; Bethlehem, a Moravian foundation turned munitions centre; York, once a grain market, includes musical instruments as well as farm machinery among its specialities. Most inland towns are also bridge towns and their growth is closely associated with the rise of valley communications.

South of the Potomac, the coastal plain broadens, woodland takes precedence over cleared land, population becomes predominantly rural, heavy industry yields to light industry, the coloured element in the population increases. The 'flatwood shore', with its small holdings, thriving peanut crops, corn, and intervening loblolly, hemlock and poplar woods, blossoms with an urban community at the broad estuary of the James River. Hampton Roads, Newport News, Norfolk and Portsmouth, linked by a bridge and tunnel across Chesapeake Bay to the Delmare peninsula, gather together commercial and naval shipping as well as exporting West Virginian coal. Inland, the piedmont zone also broadens and supports thriving mixed farms with a 9–10 month grazing period and double-cropping. An element of the sentimental persists in the landscape from the occasional plantation (more a social than an economic proposition) to the whimsy of a reconstructed Williamsburg (colonial capital of Virginia). But 'summer and smoke' begin to take control of life. The growing season extends up to 260 days and the cash crop, which is essentially a small man's crop but a skilled man's, is tobacco. The very seed is as smoke—one plant produces a million seeds, a teaspoonful of which plants ten acres. White muslin pampers the springtime seed-beds, pickers and tiers move among the big-bladed, shoulder-high plants in the fall, aluminium-roofed tobacco drying-sheds are perfumed with the precious weed and workshops, rarely employing more than several hundred, are crowded principally in Richmond. Richmond is the home of cigarettes the names of which go round the world; but 'Tobaccoland' continues south in the Carolinas and west in Kentucky. It is a paradox that this valued crop, ripening to perfection in Virginia, should also have given its name as a symbol of rural slumdom in *Tobacco Road*. Virginia is transitional from the high living standards of the north-east to the pockets of poverty that plague the South.

New England is Appalachia by the sea—Appalachia without its coastal plain, without its concealed resources of fuel, the least climatically favoured part of the province in America, that part most substantially affected by glaciation. This poorly endowed area was occupied at an early stage in American evolution by a highly selected and ingenious people. As a result, its six closely integrated states form one of the most interesting fields of human geography in the U.S.A. Bernard de Voto has called it 'a finished place'. The adjective has a positive as well as a negative ring, for he quotes Venice and Florence in the same context. There is a sense of completion about parts of New England: there are projections of Europe in it (in village form, in house shape, in church spire). It has an undeniable, even assiduously cultivated, sense of the past. But few parts of the U.S.A. have had so great a struggle to reconcile the past with the present and to arrest the farms, villages and towns that run downhill. New England grapples with the problem of industrial adjustment in its most acute form. Its economic difficulties are exaggerated because most farming in New England has long since been bypassed by more favoured lands beyond the range.

New England has always looked to the sea and the detail of its coast has guided local evolution. There are resistant rock shorelines, such as Magnolia Reef poetically associated with *The Wreck of the Hesperus*; weak rock shorelines, such as the Triassic lowlands around New Haven or the Carboniferous embayment of Boston; shorelines where glacial deposition takes precedence, e.g. Martha's Vineyard, Nantucket Island or Cape Cod; wave-built formations, such as sand reefs, tombolos, bars and lagoons. Settlement has favoured the younger rock basins pocketed in the crystalline rocks, e.g. Boston, Narragansett (Rhode Island), the Casco lowland around Portland, largest city of Maine: or the troughs etched in the interior mountains, the Connecticut 'graben', Berkshire and Vermont valleys. On its landward side, New England is cut off from the Hudson Valley by the longitudinal ranges of the Taconic, White and Green mountains, which rise to heights of several thousand feet (Figure 22). From their watershed to the seashore New England was originally clad with forests. Four fifths of Maine and two thirds of New Hampshire are still forested, though stands are of very variable quality. Timber was the richest product of the land.

Limited resources forced a series of adjustments, and the pressure,

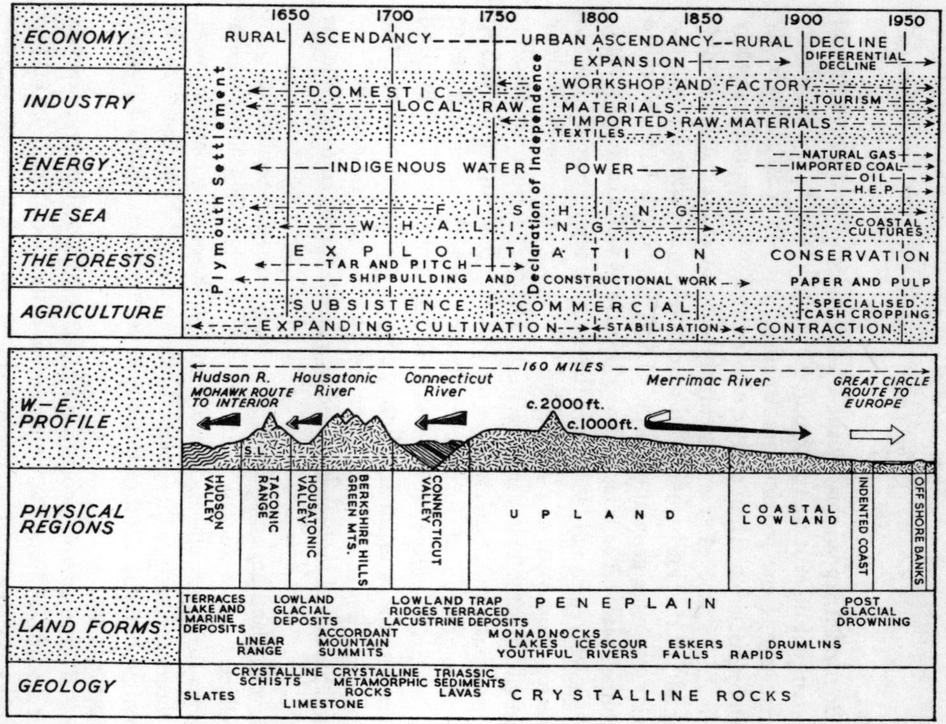

Figure 22 A Transect
through Central New England

Upper chart — time scale: 1650 1700 1750 1800 1850 1900 1950

ECONOMY: RURAL ASCENDANCY — — — URBAN ASCENDANCY — — RURAL DECLINE | DIFFERENTIAL DECLINE

INDUSTRY: DOMESTIC — — WORKSHOP AND FACTORY — — EXPANSION | DECLINE; LOCAL RAW MATERIALS / IMPORTED RAW MATERIALS; TEXTILES; TOURISM

ENERGY: INDIGENOUS WATER POWER; NATURAL GAS; IMPORTED COAL; OIL; H.E.P.

THE SEA: Plymouth Settlement; WHALING; FISHING; COASTAL CULTURES

THE FORESTS: EXPLOITATION — — — — CONSERVATION; TAR AND PITCH; SHIPBUILDING AND CONSTRUCTIONAL WORK; PAPER AND PULP

AGRICULTURE: SUBSISTENCE — — — COMMERCIAL; EXPANDING CULTIVATION — — STABILISATION — — CONTRACTION; CASH-CROPPED

declaration of independence

Lower chart:

W—E PROFILE: Hudson R. — Mohawk Route to Interior; Housatonic River; Connecticut River; 160 MILES — Merrimac River; c. 2000 ft.; c. 1000 ft.; S.L.; GREAT CIRCLE ROUTE TO EUROPE

PHYSICAL REGIONS: HUDSON VALLEY | TACONIC RANGE | HOUSATONIC VALLEY | BERKSHIRE HILLS GREEN MTS. | CONNECTICUT VALLEY | UPLAND PENEPLAIN | COASTAL LOWLAND | INDENTED COAST | OFF SHORE BANKS

LAND FORMS: TERRACES LAND AND MARINE DEPOSITS | LOWLAND GLACIAL DEPOSITS; LINEAR RANGE | ACCORDANT MOUNTAIN SUMMITS | LOWLAND TRAP RIDGES TERRACED LACUSTRINE DEPOSITS | MONADNOCKS LAKES ICE-SCOUR YOUTHFUL RIVERS | ESKERS FALLS | DRUMLINS RAPIDS | POST GLACIAL DROWNING

GEOLOGY: SLATES | CRYSTALLINE SCHISTS | CRYSTALLINE METAMORPHIC ROCKS LIMESTONE | TRIASSIC SEDIMENTS LAVAS | CRYSTALLINE ROCKS

though constantly changing in form, is persistent. They forced early and regular exchange of goods, encouraging shrewdness in trading. They invited invention and innovation, and it is logical that Connecticut should have earned the title 'the gadget state'. Mechanical ingenuity favoured the harnessing of the multiplicity of local falls, with their sites for small-scale, scattered workshops. Again, limited resources urged a search for fortune in and on the seas. Abundant hardwood and softwood, pitch and tar aided shipbuilding. And there were many protected harbours and anchorages.

Ships were built to catch cod and herring from the Banks, and to hunt whales in the North Atlantic. The Banks' fisheries are of continuing importance; the catch goes chiefly to Gloucester and Boston. They are supplemented by the rise of fish cultivation in coastal waters (oysters, clams and scallops). New Haven's and New Bedford's whaling activities, immortalized in *Moby Dick*, never really adjusted themselves to the shift to Antarctica. Ship-building skills are still employed in the yards at, for example, Quincy, south of Boston. Lesser crafts are still practised in the sail-making, roperies and cordage works of Gloucester. But it is the maritime legacy from the past which makes a greater impact—the relics of the China trade, the pictures of schooners and clippers in maritime museums such as that of Nathaniel Hawthorne's Salem, or in the homes of merchant princes as in Newburyport High Street.

Economic pressures exaggerate the physical problems of the farmer who suffers a proverbial instability of climate, long winters and unrewarding soils. Holdings are small in size and frequently awkward in shape. Competition from more favoured parts of the great free-trading area of the U.S.A. is unremitting. Factories have forced up the price of hired labour and increased its mobility. Mechanization is too expensive for the small farmer. As a result, agriculture, which diminishes in importance from rural Maine and Vermont to industrial Massachusetts and Rhode Island, employs well under a tenth of the population. More than anywhere else in the Appalachian north-east, New England has lost land through farm abandonment. Deserted holdings, with derelict farm buildings—brambles and golden rod invading their stone-walled fields—are a common country sight. Locally, there has been resuscitation. Areas of specialized farming prevail, e.g. tobacco and onions on heavily fertilized land in the Connecticut Valley; potatoes in the Aroostock Valley; cranberries in the coastal marshes of Massachusetts; poultry (cf. Rhode Island 'Reds', White Wyandottes, born of the China trade); walnut and hazelnut crops. There is also a good deal of part-time farming by the families of factory employees as well as by hobby farmers. Farmhouse accommodation for both winter and summer tourists is an important supplement to income. Forestry is inseparable from farming, but economic exploitation is frequently hampered by scattered ownership of timber stands. Large-scale afforestation, especially in Maine and New Hampshire, underwrites the softwood industries of such valleys as those of the Penobscot, Kennebec and Androscoggin.

Industry originally looked to local raw materials and local power. Much of it has a pre-steam-age location and such a watercourse as the Merrimack, with its multiplicity of manageable falls and towns (Haverhill, Laurence, Lowell and Manchester), still illustrates the localizing features. Today, import of fuel and industrial raw materials is essential. As with many industrial areas of the Old World, New England therefore stands in a distinctly disadvantageous situation. Raw materials enter principally by sea routes and the area imports six tons for every one exported. The unfavourable balance is, however, primarily a domestic matter, provides some measure of the degree of industrial refinement and reflects the native and immigrant skills concentrated in New England. Geographical inertia plays a role in the persistent location of much activity in New England, but the burden of obsolescence can be deadening.

Nowhere is the problem more depressing than in the textile industry, much of which is heavily underwritten with government orders. Cotton manufacturing began in Providence (R.I.) in the 1790's: the shift in the centre of gravity of cotton-textile manufacture to the Old South had already taken place by 1924. Silk textiles experienced a westward migration to New Jersey and Pennsylvania. Woollen textiles have suffered less displacement: the Blackstone River of Rhode Island is still lined by many prosperous mills. New England remains the leading American producer of men's suiting and such cities as Boston have large-scale ready-made clothing and hat factories. Boston is also America's leather market and it has a long tradition of shoe-making. Lynn (already producing a million pairs a year in Jefferson's day), Brockton, Salem, Newburyport and Haverhill are all shoe-making towns. Shoe-making machinery and manufacture of shoe boxes are tributary industries. The minute division of labour, the detail of distribution and the effects upon them of instability due to changes in fashion have prompted Edgar Hoover's classical industrial monograph on the industry. The manufacture of rubber footwear is a legacy of the original location of the Goodyear rubber concern in New England.

Metallurgy is one of the most prosperous branches of industry—not large-scale, heavy industry; but small-scale, light manufacturing. Electrical equipment, complemented by new electronic devices, is associated with such cities as Springfield and Worcester, also home of the American Steel and Wire Company. New England excels in a range of metallurgical products from ironmongery to precision instruments, cutlery to clocks, typewriters to small arms, jewellery to needles (including hypodermic), with a division of labour which would delight a latter-day Adam Smith.

In general, New England's industry is characterized by immense variety. Of the 350 different industrial categories listed in the U.S. industrial census, two thirds are found in the New England states. A city such as Worcester can claim a thousand different major products. At the same time, the prevalence of relatively small-scale establishments compared with the plants of the newer states is noticeable. Ultimately, however, markets must be maintained by specialist and quality products. Together with Illinois and California, the New England states continue to lead in the number of new patents granted annually. Such figures reflect the continuing tradition of innovation and the local will to maintain status.

Two other industries bear a distinct relationship to this status. The 'Student Industry' caters for the needs of tens of thousands of students in more than fifty places of higher learning. Among these are some of the most distinguished universities of the 'Ivy League' and America's best-known Institute of Technology, at Cambridge (Mass.). Important book and publishing houses are associated with this 'industry'. The tourist industry originates in both physical and human facts. The New England states have historical novelties to offer, they lie upon the threshold of the most populous and wealthy area of America, while their coastal tracts offer escape from summer heat and their northern hills provide a playground for winter sport. The coast is littered with estates and cabins, motels and camps.

Boston is the urban focus of the six states and the rest of New England's many towns and cities are as much satellites to its sun as in the time of Oliver Wendell Holmes. It is a leading port as well as a versatile manufacturing centre with 5,000 factories, but regional functions generally take precedence over national because of the long land haul. The immense and old-established diversity of shipping lines which use its deep-water harbour partly counteract locational disadvantage. In setting, Boston is an estuarine city, its harbour ringed with drumlin hillocks and the impressive Charles River shrinking inland from Cambridge. The highly congested down-town district, with its entrenched property and land rights, curiously free from a skyscraper core, mingles splendour and squalor. Boston retains an almost patrician hierarchy of leaders, whose exclusiveness tends to be exaggerated in the face of a large population of immigrant or first-generation American stock. Distinctive ethnographic distributions provide striking urban anomalies, such as the Italian occupation of much of the eighteenth-century quarter. Boston's banking institutions and merchant houses cannot change those circumstances which depress the New England economy, but their accumulated wealth and experience does much to guide the course of development in other parts of America. New England manages to acquire a full share of government contracts. Although many of their original functions are being steadily absorbed in the greater north-eastern quadrant of the U.S.A., although obsolescence and large-scale conversion may wrinkle their brows, New Englanders nevertheless remain as much controlling as controlled in the administration of the nation.

NEW YORK: THE CITY AND ITS TRIBUTARY AREA

At the meeting ground of New England and the Tidewater Lands lies the estuary of the Hudson River and around it megalopolis reaches its climax. New York City is such an overbearing phenomenon that it obscures New York State. Largely because of the city, the state's urban population (83 per cent) heavily preponderates over rural (5 per cent farm population and 12 per cent non-farm). The Hudson-Mohawk artery commands attention as a secondary phenomenon, while at the westernmost end of the Empire State, sharing the characteristics of the 'Manufacturing Belt', is Buffalo. Yet, the isthmian state of New York has a personality of its own and, as a component of the north-east, it is curiously neglected. Apart from exceptional developments along the Erie railroad, e.g. the secondary line of cities represented by Binghampton, Oswego, Corning and Elmira, population is concentrated in the lowlands. The lowlands, in turn, may be divided into the Hudson-Mohawk lands, the coastal plains of the lower Great Lakes and the valleys of the Finger Lake country.

The intervening upland masses, often rising sharply from the lowlands, were already being colonized and unwisely cleared for farms over considerable areas in the late eighteenth century. A retreat comparable to that in New England has ensued. Surveys by agriculturalists from Cornell have revealed that the number of farms in the state fell by a third between 1900 and 1940: in some counties, by a half. Amalgamation partly explains this; but retreat of the frontier of occupation as well as that of cultivation has been widespread. Farmland has been extensively restored to forest through central agencies. Simultaneously, tourists have multiplied in this unexpectedly attractive part of the country, while in such cities as Saratoga Springs or Lake Pleasant the Adirondacks have summer spas to balance winter ski resorts.

Agriculturally, New York is a dairying state, and the demand of the urban population for fresh milk gives to it a production status following upon that of Minnesota and Wisconsin. The warm valleys of the Finger Lakes Seneca, Cayuga, Owasco, Skaneateles, Canandaiga, enriched by soils high in lime content and humus, provide the state with fruitful basins which in their production recall those of the Pacific coast. Largely as a result of them, New York State has an apple harvest second only to that of Washington, a grape harvest second only to that of California (albeit a long way behind) and a tobacco output larger than that of Virginia (though coarse in quality).

The lowland farmsteads frequently lie near industrial towns, of which there are a full fifty on the 400-mile railroad stretch from Buffalo to New York City. Industry closely resembles that of New England and is frequently of New England parentage. It is heavily dependent upon imported raw materials, of a light manufacturing character and highly specialized. Illustrative of the consumer goods derived from a tract of cities—the juxtaposition of whose names is an Old World toponymical nightmare— are brass and copper products (from Rome), typewriters (from Ilion), shirts and collars (from Troy), carpets and rugs (from Amsterdam), soda and chemicals (from Syracuse, which has brine deposits) and electrical and metallurgical manufactures (from Schenectady and Albany, the state capital). Rochester, on the south shore of Lake Ontario, has one of the best-balanced industrial structures of any town in the industrial north-east, including semi-luxury products (e.g. optical instruments and cameras). Power comes from coal (near at hand), natural gas (on the spot) and hydro-electric power (local small-scale), and St. Lawrence River (large-scale).

Most of the area is tributary, directly or indirectly, to New York City. Descriptions of this polyglot urban complex of fully 10,000,000 inhabitants, where one in every four is of foreign origin, where a birth takes place every four minutes and a marriage every seven, usually dissolve in superlatives. In the context of the north-east quadrant, however, there are several features that merit emphasis at any level. In the background to New York City are gathered together the features both of New England and the Tidewater Country. At the same time these features surround the gateway to the continent. No city on the Atlantic seaboard experiences the structural diversity of Appalachia more fully and none is more sensitive to the interplay of land and water.

As can be seen from Figure 24, greater New York has an insular and peninsular setting. It is centred on three main islands (Manhattan Island, Staten Island and Long Island) and two peninsulas. From the seaward side, the main approach is through the 40-foot-deep Ambrose Channel between

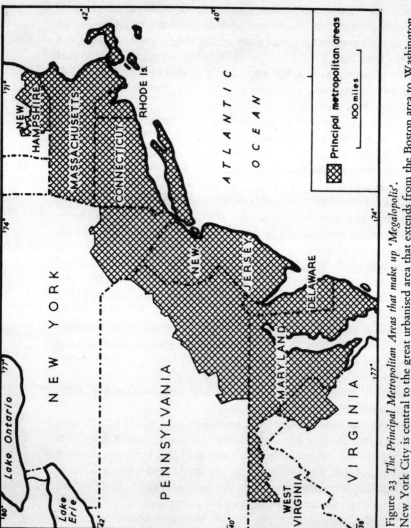

Figure 23 *The Principal Metropolitan Areas that make up 'Megalopolis'.*
New York City is central to the great urbanised area that extends from the Boston area to Washington D.C. 'The Main Street of the Nation' is the epithet used for it by Jean Gottman in his examination of its characteristics and problems. (Source: *Megalopolis*, New York, 1961)

the spits of Rockaway Beach and Sandy Hook, to the Lower Bay and thence through 'The Narrows' to the Upper Bay. A second approach is by way of Long Island Sound (with a 20-foot channel) to the East River by central Manhattan. Long Island and Staten Island, detached eastern extensions of the coastal plain, are built upon Cretaceous foundations though their relief is determined principally by overlying morainic materials. Their low-lying and easily eroded shores include extensive swampland. Manhattan Island and the adjacent Bronx are crystalline projections from New England. The New Jersey extensions of Greater New York rest upon Triassic sandstones comparable to those of such eroded areas as the Annapolis-Cornwall Valley of Nova Scotia. But their smooth relief is interrupted by igneous traps which flank the right bank of the Hudson as the upstanding Palisades. Successive intrusions, following the north-south axis of the Hudson, are repeated westwards, the most prominent being the Wachtung Range. Beyond Jersey City lies Newark Bay, with its tributary Hackensack and Passaic rivers.

The built-up area (Figure 24) has its own particular responses to these varying bedrocks, while mercantile life is sensitive to the form and accessibility of the coastline. Thus, the crystalline rocks of Manhattan Island are a firm foundation for the *horst* and *graben* of its skyscrapers, buildings in which considerations of prestige probably mean more than economy on ground space. Contrastingly, the hard rock base has hampered tunnelling and underground communication; while the form of the Palisades has restricted building development. Harbours, wharves and coastal installations are most continuous along the hard rock coasts of Manhattan Island and the Bronx, which are encased in quays and piers. Tidal range is only 4 or 5 feet in these inner reaches and docks are unnecessary. Reclamation of low-lying tracts has been encouraged by increasing pressure upon the land. The flatlands of Long Island have been drained for airports, e.g. Idlewild, La Guardia by Flushing Meadows; Floyd Bennet Field at Rockaway Inlet. Hackensack Meadows and Jamaica Bay still have extensive unreclaimed areas.

Ease of movement is a primary concern of any big city. New York's amphibious setting has called for a network of rail, vehicle and passenger ferries; but bridges and tunnels, though expensive to construct and maintain, now offer swifter routes. The port of New York also relies on some 100,000 barges for the movement and transhipment of cargoes. The transport economics of the city are very much three-dimensional. Vertical as well as horizontal movement is important not only for office workers and daministrators but also for downtown workshops. The clothing industry, for example, concentrated in Lower Manhattan and probably contributing the largest single proportion of ready-made clothes to the female market, operates from sky-high premises.

New York City, made and marred by waterways, has had to wrestle with its water supply. Only during the last decade has the city's water supply been placed upon a secure foundation. The hill country to north-east and north-west has an abundance of lakes; but despite elaborate inter-linking they proved inadequate. New York City, now fed by aqueducts which stretch as much as 150 miles into its hinterland, is sustained by major dams. Sewage disposal has presented a complementary problem, because of the feeble tidal scour.

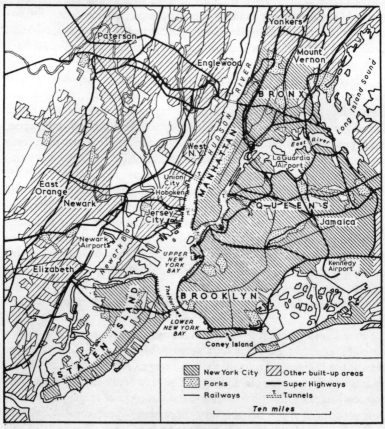

Figure 24 *New York City*

As in all big cities, New York has its regional specializations: as with all American cities, it has its ethnographic differentiation. Harlem's 50-year-old association with the Negro community is world-renowned. Industries are strongly consumer-oriented and the percentage of luxury trades, such as jewellery and furs, is high. There is also specialization in trading facilities. Dock specialization, for example, is closely related to adjacent landform and land use. The larger and heavier traffic is generally directed to island shores: the lighter, to peninsular shores. Thus, West Manhattan receives the largest passenger liners; while Staten Island is the terminal for bulk cargoes, such as crude oil, coal, ores, timber, clay and sand.

A hundred and fifty years ago, with its 100,000 population, New York City was mappable and negotiable. Today it defies geographical treatment. A geography of the city would be more formidable to compile than a geography of London. In the first place, the scale of its operations is so much greater. The port, for example, receives a greater tonnage than that entering the port of London, and its passenger terminals at the heart of the city are as if the functions of Southampton were added to those of London. The value of its overseas trade is even more impressive than its volume. It accounts for nearly half of America's overseas trade by value: in addition, it has an important cabotage. In the second place, New York is difficult to assess geographically because of the speed of change within it. Its population is mobile: its urban features subject to replacement with a frequency unknown in the Old World. Finally, New York is a city of world consequence as well as of national import. The world is sensitive to Wall Street, a foreign consul in New York City counts for more than an ambassador in a good many capitals, the very fate of the world is debated in the U.N. building which looks out upon the East River. New York City, it has been fairly said, 'breathes of things too large for books'!

THE POWER OF THE HINTERLAND

The basis of the widespread and intensive industrialization of the north-eastern U.S.A. has been coal. Appalachia continues to produce more than two thirds of America's 400,000,000 short tons annual output and most of it derives from the bituminous deposits of Pennsylvania and West Virginia. The coalfield is virtually coextensive with the Appalachian plateau. On its northern margins the more exposed portions have suffered erosion: on its eastern front there has been folding. Coal is located in three main areas—in Pennsylvania around Connellsville, in northern West

Virginia 150 miles south of Connellsville, in southern West Virginia around the New River. West Virginian production now takes precedence over Pennsylvanian. Eastern Pennsylvania has anthracite in the Susquehanna and its tributary valleys.

Coal deposits are characterized by accessibility. Seams are but little folded, many are thicker than 6 feet. The plateau area displays pits of all sizes and in all stages of evolution. Most coal is mined, but there is some open-cast working; while occasional back-garden pits suggest the Forest of Dean in their sylvan setting. The deeply incised drainage pattern contributes to accessibility. Most seams lie about 400 feet below the surface. In the north, the Allegheny, Youhiogeny and Monongahela are incised about 500 feet; in the south, rivers are entrenched about 1,500 feet. Adit mining has been employed with great effect and, until exhaustion of seams, upslope galleries facilitated drainage as well as gravity movement of trucks. Mining settlements have tended to concentrate in the valleys along the lines of canalized river transport. Feeder rail tracks have struck back into the hills. While movement downstream has been aided by waterways, transverse contacts have been much hampered by entrenchment, especially in West Virginia. Within the area coal travels by water; but most of that which leaves the area moves by rail. About equal amounts move Atlanticwards and to the Great Lakes cities. In few mining areas has mechanization proceeded more swiftly. The number of miners is correspondingly small, some 200,000 with an output of 11 tons of coal per head daily. As labour required has become increasingly skilled, it has become correspondingly less mobile.

Beyond the Allegheny front, anthracite is found in a concentrated area of specialist production largely in the Lackawanna Valley. The seams are contorted, compressed, frequently fractured, several thousands of feet below ground level, shaft-mined and not easily mechanized. Costs of production are higher, output per capita lower than in the coalfields and production (c. 20,000,000 shaort tons) is declining. Carbondale, Scranton Pottstown, Wilkes-Barre nd Nanticoke form an interconnected urban, area to which the anthracite area is tributary. On the flanking hills, coalpits, sometimes open-cast, look down to the newer-planned anthracite towns of the valley bottom (Figure 25). Appalachian woodlands conceal much of the industrial landscape and are continued in the tree-lined avenues of Scranton. The employment columns of the local press reflect the stagnation of mining. Scranton lives by its clothing and tobacco industries, originally attracted here by the abundance of female labour.

Coal and Appalachia go together, but the plateaux have in fact a trinity of combustible fuels. Oil and natural-gas occurrences are wide-spread, but yields of oil are capricious. The first spouting wells were tapped in Pennsylvania and the quality of its oil is high. Volume is small and meets competition from oil imported by trans-continental pipe-lines. Natural gas is now much more widely used than coal gas in a state which a generation ago had one of the world's most active coking regions. No

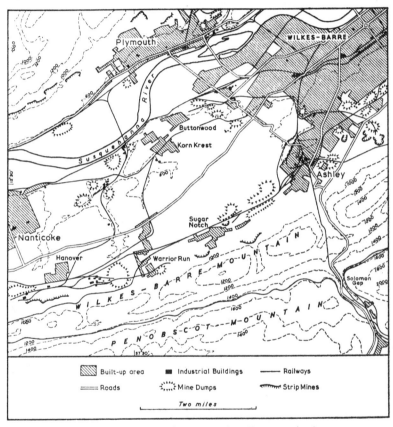

Figure 25 *The Landscape around Wilkes-Barre (Pa.)*

In the upper reaches of the Susquehanna Valley, mining and industrial communities have grown side by side for the better part of a century. The entrenched river has exposed coal seams which are strip-mined on the Appalachian flanks. Urban settlements occupy the flood plain and the entire landscape is heavily wooded. (Source: *U.S.G.S.*, 1 : 25,000, Wilkes-Barre West Quadrangle)

part of the U.S.A. has a more intensive network of natural-gas pipe-lines than Pennsylvania and Ohio.

This concentration of energy resources in the hinterland of the north-eastern seaboard has been fundamental to the emergence of the 'Manu-facturing Belt' (Figures 26, 27 and 28). From the end of the eighteenth century onwards, the Pennsylvanian hinterland of the Tidewater Country saw the spread of iron and steel manufacture. Valley Forge (a place-name made famous by George Washington in other ways) is a symbol of many small sites that combined local charcoal, ore and water power. The even-tual focus of the industry, striking a liaison between excellent coking coal and imported Superior iron ores, was Pittsburgh. It remains a pivot of a steel industry, the magnitude of which must be viewed in terms of the European Iron and Steel Community.

A galaxy of not less than forty towns, with populations of 10,000 and upwards, strung principally along the river axes, has grown up around it. They draw their materials from it, are mostly concerned with secondary and tertiary processing, and produce such goods as plates, wires, sheets, tubes and structural steels. A second group of cities and towns extends north-westwards through Youngstown and Canton along the shortest and easiest connecting route between Pittsburgh and the lakeshore. North-eastwards, Buffalo is a metallurgical outlier and beyond it are located the mills of Hamilton (Ont.) and the processing plants of the St. Cather-ine's area. South-westwards, the limit of steel processing is at the head of deep-water navigation on the Ohio at Wheeling.

Transport costs are fundamental to an understanding of this distribu-tion. There is first the comparative cost of rail transport which is about ten times as much as water transport. Secondly, there is the proportion in which different raw materials are combined. Thus, the volume of coking coal needed to smelt a ton of iron ore has historically favoured the move-ment of ore to coal, though the proportions change with changing techniques. Movement over natural water routes has been complemented by movement over improved waterways. Canals never achieved any importance in the iron and steel complex, but the canalized reaches of the upper Ohio provide an important integrating network. Winter inter-ruption on the Great Lakes calls for stockpiling and also imposes shipping risks, especially at the close of the season. In the shipment of Lake Superior ores to the Pittsburgh area there was an automatic challenge to its supremacy, because return coal cargoes favoured the Erie coast for alter-native sites.

Settlements on the south shore of Lake Erie have responded sensitively

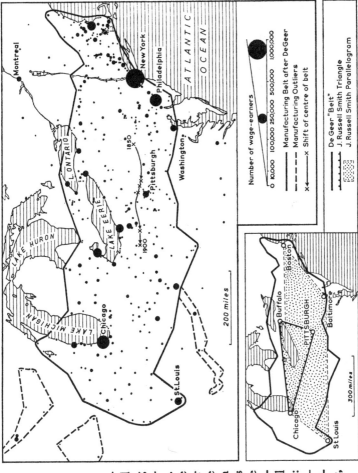

Figure 26 *The Manufacturing Belt of the North-Eastern States, after Sten de Geer*

The Swedish economic geographer, Sten de Geer, identified the north-eastern manufacturing area according to the distribution of industrial wage-earners. His study was based upon the 1919 census. He took as his lower limit for representation the unit of 1,000 wage-earners. In casting a boundary round his manufacturing belt, he chose empirically a distance not exceeding 53 miles between individual manufacturing units. (Source: *Geografiska Annaler*, 1927, 233–59, and J. Russell Smith, *Industrial and Commercial Geography*, New York, 1913)

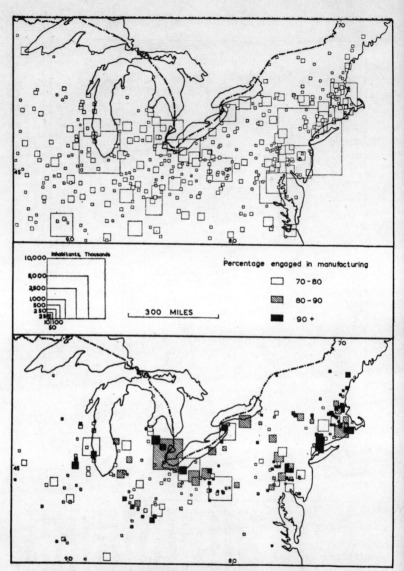

Figure 27 *An Interpretation of the North-Eastern Industrial Area, after Gunnar Alexandersson*

Gunnar Alexandersson has mapped the distribution of American cities with 10,000 inhabitants or more in 1950, and classified them according to the percentage of people engaged in manufacturing (Source: *The Industrial Structure of American Cities*, London, 1956, reproduced by courtesy of the author)

to their intermediary location as receiving and dispatching centres of primary raw materials. Cleveland rapidly emerged as an alternative site for steel processing. Ashtabula and Erie to the east and Toledo to the west have been drawn into the orbit. Toledo and Ashtabula have specialized in the bulk handling of coal. Toledo claims to be one of the largest coal-exporting harbours in the world. Detroit has blast furnaces associated with its automobile industry. The most striking reaction to

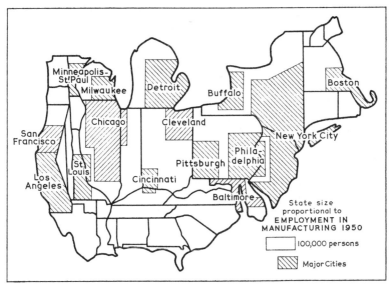

Figure 28 *A Manufacturing Map of the U.S.A., by Chauncy D. Harris*

In this map, based on 1951 censal material, an attempt is made to express the states of the union in their relative size according to manufacturing output (Source: 'Market as a Factor in Location of Industry', *A.A.A.G.* 44, 1954, reproduced by courtesy of the author)

changes in the economics of production has been in the rise of iron and steel plants at the south end of Lake Michigan, where Calumet had its first workshop in 1880. Limestone for fluxing is accessible to almost all processing sites. Scrap iron, of which the U.S. steel industry consumes a full 50,000,000 tons annually, is also abundant in the north-east.

Market control confirms the location of the industry in the north-east. Distribution costs are generally high, so that producing units are most efficiently located near to the consuming area. In the north-east the source of fuel and the main consuming area overlap. But although the industry may have a strong location for the national market, its 'inland' location is a disadvantage for the international market. The opening of the St. Lawrence Seaway strengthens the position of the lakeshore sites; the more so, with the rise of the Labrador ore supplies. The expansion of plants on the Atlantic foreland is also a likely response.

In the modern phase of the industry's evolution, two related changes may be seen at work, one structural and the other distributional. Both raise difficulties for Pittsburgh. Structural changes are seen in the rise of large integrated mills, embracing blast furnaces, steel mills, rolling mills and specialized processing units. These in turn require spacious sites. A glance at the topographical map of the Pittsburgh area shows a complete occupation of the level riverside terraces suitable for new plants. Many firms have therefore opened new branches near to the Erie shore, where suitable sites for large factories are available, and the tendency is to transfer attention to them as inland plants become obsolescent.

The effects of iron and steel processing—the move away from its primary inland centres—have been exaggerated by the absolute decline of the coking industry. The Connellsville area has suffered most, for a generation ago roughly half of America's coking ovens were concentrated around it. Nor has the reappraisal of by-products such as coal tar, ammonia and benzol derivatives helped greatly, for their processing plants tend to be located not at the site of raw material production, but where the market is greatest. The chemical industry has developed more vigorously in West Virginia than around the older coalfields.

Pittsburgh, the city of Andrew Carnegie whose company became central to the United States Steel Corporation 60 years ago, at that time held a near monopoly position in price control. Within 60 miles of the city some 40 per cent of America's steel was produced. In 1900 a practice of pricing iron and steel products according to the price prevailing in Pittsburgh plus the cost of transport from Pittsburgh to the market was established, wherever the market might be. This price control lasted a generation. There is little in the external appearance of the city to suggest decline. Despite hilly environs that give it gradient problems even worse than those of the steel capital of Britain, Pittsburgh's urban area is extending as swiftly as that of most American cities. Its downtown section, with gleaming aluminium-faced skyscrapers at the Fork of the Ohio, has undergone a more fundamental transformation than almost any city of the same age. The pall of smog which used to fill its threefold valleys has been eliminated. And if traffic diminishes on the nine trans-Allegheny railroads that converge upon the industrial capital of Appalachia, it mounts ceaselessly upon the Pennsylvanian turnpike which ties it to Philadelphia. Pittsburgh has relaxed, but not released its control of steel. It has yet to be said of any U.S. steel city that it is 'Pittsburgh plus'.

'JE MAINTIENDRAI'

The motto of the House of Orange, which must have been familiar to New Amsterdam, would not be inappropriate to a crest for the North-East Atlantic States. In many fields their task is to resolve the problem of maintenance. They were in the vanguard of development. For them, it is less easy to replace the old with the new, or to combine the old and the new, than it is for younger states in the west to begin their development without trial and error. Fundamental shifts in status affect the north-east as a result of changing demand and the employment of new raw materials.

Coal and anthracite, for example, remain essential fuels for the daily existence of the nation; but they only account for about a third of its energy requirements as distinct from nine tenths 60 years ago. They have yielded much to petroleum and to natural gas. The situation is exaggerated by the immenence of a major national breakthrough in the production and use of atomic energy. The pockets of poverty in the mill towns of Lawrence or Fall River are repeated in the coal towns of Appalachia, where three out of five miners may have been affected. In the field of steel production, the absolute status of the north-east is unchallenged, yet South Chicago now claims the largest regional output of any steel-producing complex in America.

Demographic statistics provide another example of changing status. If numbers mean anything New York State, with the largest population, must be the most desirable state in which to live. Yet California, 'The New Empire State', now threatens to take the lead over the old 'Empire State'. There are also changes in trading status. New York City remains unrivalled as America's leading port, but other ports of the Atlantic seaboard have forfeited pride of place to Houston, Texas.

Industry began in the north-east and the means of control remain there, from the steelmasters of Pittsburgh, through the merchants and moguls of Boston or Baltimore, to the bankers of New York and the administrators of Washington. In the final instance it is the changing emphasis on individual branches of industry that causes the north-east to feel the winds of change. The growth of other industrial centres makes a greater impact on the casual observer than the continuing growth within the established north-east. The U.S.A. can continue to maintain the old-established manufacturing belt as well as to support the growth of other industrial areas.

SELECTED READING

The broader panorama of the industrial north-east has offered an almost too strong challenge to geographers. Beside Sten de Geer, this challenge has been accepted by

Jean Gottman, *Megalopolis* (New York, 1961)

For the reverse of the coin, there is.

L. Klimm, 'The Empty Areas of the North-east U.S.A.', *G.R.* 44, 3, 305–45
Outdoor Recreation for America (Washington, 1962)

See also

C. D. Harris, 'Market as a Factor in Localization of Industry', *A.A.A.G.* 44 (1945), 315–48
W. Isard and W. M. Capron, 'The Future Locational Pattern of Iron and Steel Production in the U.S.A.', *Journal of Political Economy*, 17 (1949), 118–33
R. C. Estall, *New England, a Study in Industrial Adjustment* (London, 1966)

The problems of New England have sponsored many investigations, e.g.

J. D. Black, *Rural Economy of New England* (Cambridge, 1940)
E. Hoover, *The Boot and Shoe Industry of New England* (Cambridge, 1940)
E. Ackerman, *The New England Fishing Industry* (Chicago, 1941)

It has also been a fruitful territory for studies in sequent occupance, e.g.

P. E. James, 'The Blackstone Valley', *A.A.A.G.* 19 (1929), 67–109
E. Ackerman, 'Sequent occupance of a Boston suburban community', *E.G.* (1941), 61–74
H. M. Mayer and C. F. Kohn, *Readings in Urban Geography* (Chicago, 1959)—includes relevant chapters
M. J. Proudfoot, 'Public Regulation of Urban Development in the U.S.A.', *G.R.* 44, 3, 415–19—adds an important note on urban controls

Other useful texts include

R. C. Estall, 'Appalachia State', *Geography*, 53 (1968), 1–24
G. F. Deary and P. R. Griess, 'Effects of a declining mining economy on the Pennsylvanian anthracite region', *A.A.A.G.*, 55 (1965), 239–59
A. von Burkelow, 'New York City's Water Supply', *G.R.* 49, 3, 369–86 (1959) —puts a vital natural resource into its proper perspective. *Made in New York* (Cambridge, 1959) explores four aspects of the city's industrial life
G. F. Deahy and P. R. Griess, 'Geographical Significance of Recent Changes in Mining in the Bituminous Coalfields of Pennsylvania', *E.G.* 33 (1959), 283–98

5

The Middle West and the Search for a Balanced Region

A Land of Mobility—The Character of the Land and the Form of the Settlement—The Economy of the Middle West—Crossroads and Crux of the Continent.

A LAND OF MOBILITY

The Middle West has been defined by many different writers in many different ways; but for the purpose of this chapter it may be assumed to consist of the area which lies between the Great Lakes to the north and the Ohio River to the south, the lower slopes of the Allegheny plateau to the east and the Missouri River to the west. It focuses on the states of Iowa, Illinois, Indiana and Ohio in the south, and Michigan, Wisconsin and Minnesota in the north. There are features identifiable as 'Middle Western' which lie outside the area, e.g. in Missouri. There are features alien to the Middle Western scene which are incorporated in the area defined, e.g. the range country of Minnesota. Some geographers, such as O. E. Baker, have located distinctive crop regions within the Middle West; others, such as Sten de Geer, have observed the thrust of the North-Eastern Manufacturing Belt into it. It is a land rich in both farm and factory: a good land, in which hog and hominy have been and are regarded as reasonable substitutes for milk and honey (though there is plenty of milk and honey as well). Canada has a continuation of the Middle West in the Lakes Peninsula, but no counterpart to it. The Lakes Peninsula resembles it in economic structure, but the scale of its operations is small by comparison. The Middle West, despite spaciousness, has a curious compactness. It is a traditional land of transit, yet it hangs together, works together and not uncommonly thinks together.

From the first, the Middle West and its surrounding territories have been lands of mobility (Figure 29). In this country of low and sometimes diffuse watersheds neither landform nor waterway has offered serious obstruction. The original vegetation of deciduous groveland and intervening long-grass summer prairie also imposed little restraint. The area is integrated through rather than united about its waterways. Rivers aided and continue to aid movement in it, but they direct attention beyond it. The Great Lakes offer routes to complementary shores at least as much as they bind its components together. The waterways are of diverse form, having only one fact in common—that the present volume of water is smaller than that of the glacial melt-water which originally carved their valleys or filled their depressions.

The Mississippi originates in the area, though precisely where is a matter of dispute. It is controlled by dams below the St. Anthony Falls at Minneapolis, where a long-awaited flight of locks is to link the short navigable section to the north. Its valley, sometimes several miles broad between the pyramidal bluffs of its sedimentary flanks, has pancake-flat islands, sand spreads and mudspreads where muskrats thrive, and where generations of Tom Sawyers and Huckleberry Finns have their playground. The chief east-bank tributary, the Ohio, is still a great corridor of movement. Canalized and but little affected by winter icing, it carries hundreds of thousands of passengers and tens of millions of tons of cargo annually. Middle West rivers such as the Illinois and Wabash are less well known by name, but they have played important historical roles in settlement and communication.

North of low watersheds, which raised few portage problems in earlier times, are the Great Lakes. They provide another great series of highways for the Middle West. Movement upon them is beset by the hazard of storms and by winter icing. Coastal accessibility is limited for larger vessels by wave-cut features, wave-born deposits and occasional tracts of impressive dunes, as in south-east Michigan. Mobility upon the Great Lakes has been steadily strengthened by the improvement of their constrictions (e.g. at Sault St. Marie, the St. Clair river shallows), by the construction and dredging of harbours (no Great Lakes port has a 27-foot channel) and by the growth and improvement in size and speed of vessels. 'The Log of the Lakes' indicates three dozen or more major vessels to be a fairly common day's procession through the canal at the 'Soo'. The Middle West has also many thousands of lesser lakes especially in its northern half. The 'land of sky-blue waters' has acquired a new status in the modern age of pleasure boating.

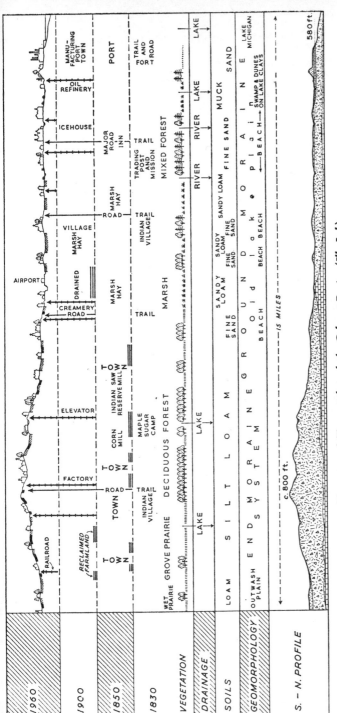

Figure 29 *A Transect through the Calumet Region (Ill.–Ind.)*

Among American geographers who have compiled transect diagrams to illustrate successive changes in landscape evolution, Alfred H. Meyer is renowned for his definitive study of the Kankakee marsh. (This transect is based, with the author's permission, upon his diagram in 'Circulation and Settlement patterns of the Calumet–South Chicago region of Northwest Indiana and Northeast Illinois', *Proc. 17th International Congress Washington, 1952* (1957) 538–44, fig. 2.) In 1960 black shading indicates grain crops and oblique lining grass crops.

While water has aided movement, the land has rarely obstructed it. The Middle West is a country of modest variations in altitude. There are the swells and swales of moraine or drumlin country, there are river-valley bluffs, but no major barriers—at least in the south. For the most part, too, it is well-drained country; though ill-drained areas, such as the Black Swamp of the Maumee River, have left their mark upon local settlement. There are generous deposits of road metals. It was in Ohio that the pattern of township and range was first laid down. There was little to prevent conversion of their concession roads into four- or six-lane inter-state motorways.

The Middle West is also natural railroad land and, if Chicago is one of the world's largest railroad transfer points, St. Louis and Minneapolis/St. Paul are no mean rivals. Diesel oil has supplanted coal as the motive power. The capacity of diesel locomotives to travel 3,000–4,000 miles without refuelling or to pull a mile or more of box cars at 50 m.p.h. is an impressive transport feat. Fat railroad timetables still give the illusion of the line of steel as a primary integrating force, while a measure of romance is still conveyed by the names given to long-distance expresses such as the Hummingbird (which runs between Chicago and New Orleans).

Yet rail traffic suffers severely from the competition of road transport. Long-distance trucking is a common feature of goods transport and the Middle West has its fair share of 100-foot-long vehicles, air-conditioned if need be. It may have been an accident that the Middle West gave birth to the first large-scale manufacture of cheap automobiles. It is no accident that the principal centres of manufacture have persisted in this land of easy mobility. The automobile is of profound social as well as economic importance, for it assists American neighbourliness. It also makes its own particular contributions to the townscape. The immoderate extent of the Middle Western city reflects the ease of personal movement. Downtown, the parking lot and underground garage are the complements of the skyscraper; out-of-town, supermarkets and shopping centres are designed as an escape for the driver from downtown congestion. Unusual urban-rural migrations have been promoted by the automobile, not least that of the suitcase farmer, who prefers to live in town and to drive out to work on the land. The car has also given rise to a new breed of nomads who live in mobile trailers.

There is lively movement above the land as well as upon it, for both the Middle Western and the Great Plains are good flying country. They abound in satisfactory airport sites and offer few climatic problems. For many, the aircraft has completely replaced terrestrial travel. The intensity of domestic airway services can be greater in few places than such cities as Chicago or St. Louis, with passenger lines leaving for the major east- and west-coast cities every hour of the day and night. Some measure of the intensity of American air services is evidenced by the fact that, in rank, Little Rock (Arkansas) is an airport of international stature. Private aircraft complement commercial. Such cities as Chicago, Detroit and Minneapolis are ringed by small private airfields, with scores of light aircraft moored against the wind.

The Middle West also permits the ready transport of energy. Electric power lines swing freely across it to clutter much of the urban as well as rural skyline. Natural gas and oil pipe-lines (some overground, some underground) criss-cross the countryside. Coal moves primarily by rail, e.g. the axial line from West Virginia to south shore of Lake Michigan. The Middle West also claims the longest conveyor belts in the world. Although the area has energy resources of its own, not least the coal reserves of the mid-continental field, its contemporary supplies are drawn principally from outside.

THE CHARACTER OF THE LAND AND THE FORM OF THE SETTLEMENT

Permanent occupation of the Middle West was relatively delayed; but in few parts of the world has it brought about such a thorough-going change in so short a time. The result is a powerfully humanized landscape, in which intensive but well-distributed settlement maximizes the oppor- tunities of unexpected diversity in natural setting.

Marked variations in relief may be absent, but great varieties of surface deposits mould and modify local opportunity. Diversity of soils persists, changed though they may be by increasingly vigorous programmes of drainage and fertilization. The motorist on a day's journey of 300–400 miles between motels in Indiana and Iowa might identify sandy moraines, boulder clays, gravel spreads, peatlands, alluviums and loess (into which the Iowan road may be entrenched 20 feet). These may be locally redis- tributed by water or wind. A traverse from Chicago to Minneapolis will carry the traveller through the remarkable driftless area, with its castellated sandstone residuals, and across deeply entrenched river valleys such as the St. Clair where windows are opened on the underlying limestones. These limestones have made their contribution to soil fertility, while vegetation has contributed a substantial humus component. Soils are predominantly

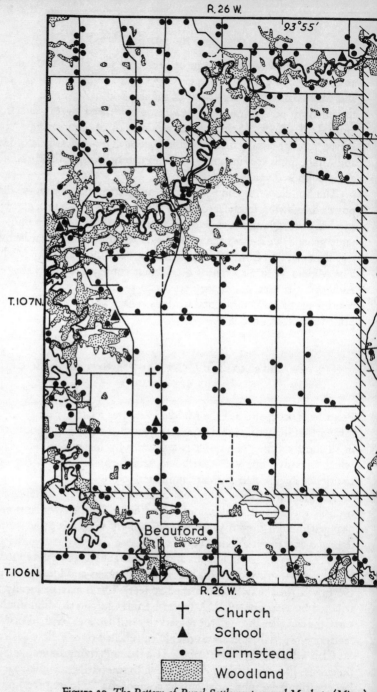

Figure 30 *The Pattern of Rural Settlement around Mankato (Minn.)*

In this representative settlement pattern from south-central Minnesota, farmsteads and rural service organizations are dispersed uniformly over a countryside in which till plains are dissected by

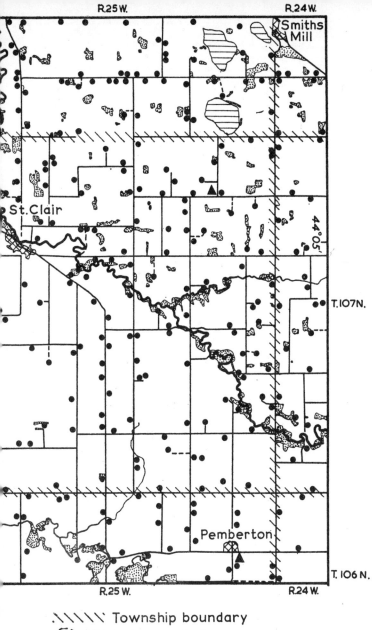

R.25 W.   R.24 W.

Smiths Mill

St.Clair

44° 05'

T. 107 N.

T. 106 N.

Pemberton

R.25 W.   R.24 W.

\\\\\\ Township boundary

Lakes and rivers

——— Road

THREE MILES

rs which bite into the underlying limestones. Mankato is the
rket and educational centre of a tract largely settled in the 1860's
(Source: *U.S.G.S.*, 1 : 62,500, Mankato East Quadrangle)

grey-brown podsols—'rich as a dunghill', wrote the forthright William Cobbett. Most of them provide a good seed bed—'only to be tickled with a hoe to laugh with a harvest'. Soil variation is as apparent on the individual farm as elsewhere, especially when turned up at the spring ploughing. The generally light and porous soils are highly susceptible to erosion, so that a main farming task is to keep the soil covered. Precipitation, falling as short, sharp convectional thunderstorms in summer, exaggerates this susceptibility. Despite its modest relief, more than half of the Middle West has slope conditions which may suffer erosion. Equinoctial winds also erode exposed soils. The southern fringes of the area have suffered relatively more than other areas, and along the Ohio and Missouri valleys the complement of erosion is river-bed silting and flooding.

Over the greater part of the Middle West the original settlement features are strongly persistent. The isolated farm predominates, while such village concentrations as existed have usually withered. In Illinois, Indiana, Ohio, a representative farm may be 150–200 acres in size and almost entirely under the plough. Fully two thirds will bear grain or fodder crops, with corn generally taking precedence. The rest will be under temporary pastures for cattle or hogs. The woodlot is usually reduced to a grove of trees. The central complex of buildings is commonly close to the road, the surface of which reflects the wealth of the community and the severity of the winter season. Buildings are almost always made of wood, with shingle or composition roofing. They are big and usually well painted, commonly in green and white. The barn (or 'cattle castle'), with its tall aluminium-capped silo tower and an upper storey containing dry fodder, houses all farm animals in winter. Corn bins may be alongside. The farmhouse, with fly screens in summer and storm windows in winter, invariably has a shaded porch or verandah. Electricity and telephone cables loop their way to it, an aluminium gas container sits at the back door, while diesel or electric machines pump water from a well that is rarely more than 20–30 feet deep.

The even distribution of farmsteads (Figure 30) is repeated in the distribution of the second element in the hierarchy, the county towns or 'cities'. They are commonly located at 5- or 6-mile intervals from each other, with populations of several hundred rather than several thousand. The county town will be constructed chiefly of wood, with the lumber merchant frequently owning its largest single business (Figure 31). The older buildings, town hall and fire hall may be of brick; while the bank may be faced with stucco or stone. The churches, often spired, announce their presence above the tree-lined streets together with the bulbous,

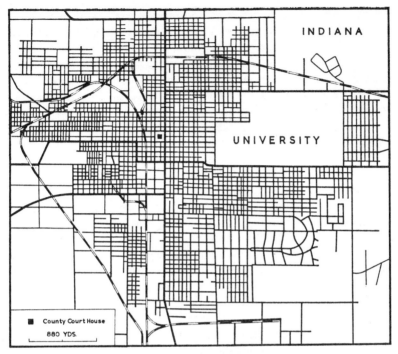

INDIANA

UNIVERSITY

■ County Court House
880 YDS.

Figure 31 *Bloomington (Ind.), a Middle West County Town*
Indiana was being staked out in farming sections during the earlier years
of the nineteenth century, and county towns were being formally laid out
simultaneously. The grid-iron pattern of tree-lined´streets, with squares commanded by court houses, administrative and ecclesiastical buildings, is
common to most county towns (after the map published by Bloomington
Chamber of Commerce)

aluminium painted water containers. The schools, almost always new and
well designed, will be filled partly with the aid of a network of school buses.
Other service institutions may be the feed mill (by the railroad, when it
exists), co-operative creamery, well-drilling workshops, garage and
machine repair shops (with a yard filled with second-hand machines),
privately owned stores, canning plants, taverns and eating houses. The
entire complex is set against a grid-iron pattern of streets, a good many of
which will be flanked by the villas of retired farmers.

The Middle West has a substantial number of cities of several tens or
scores of thousand inhabitants. Many are the equivalent of the regional

market centres of western Europe and exist essentially to meet the needs of the local farming community. Their industries are varied, with work-shops taking precedence over factories, and secondary or assembly in-dustries balancing processing of farm produce. Some cities owe their ex-istence to local ingenuity, e.g. Battle Creek (Mich.), the original home of Kellogg breakfast foods, and Grand Rapids (Mich.), one of the country's leading furniture-manufacturing cities. Some capitalize upon local re-source, e.g. Decatur (Ill.), the self-styled 'Soy Bean Capital of the World'. Some owe their substantial size to educational institutions, e.g. Blooming-ton (Ind.) and Valparaiso (Ind.). Some are specialized harbours, e.g. Ashtabula (Ohio), a coal port of Lake Erie, or Menominee (Mich.) on Lake Huron and Two Harbours (Minn.) on north shore, Lake Superior, which export iron ore. There are also the state capitals, e.g. Springfield (Ill.), Des Moines (Iowa) and Madison (Wis.). The frequency of occurrence of cities of this size is more marked in the Middle West than in most parts of the U.S.A. They proliferate upon the eastern margins, which are at the same time the western margins of the 'North-Eastern Manufacturing Belt'.

Most of the leading cities of the Middle West have waterside locations. There are lakeside cities—Chicago (with a population of several million depending upon how it is defined) and Milwaukee to its north; Detroit on the narrow channel which leads from Lake St. Clair to Lake Erie; Cleve-land and Toledo on the south shore of Lake Erie, both tied by a procession of summer ships to the twin cities of Duluth/Superior at the head of Lake Superior. There are the cities on the big rivers—St. Louis near the confluence of the Missouri and the Mississippi; the twin cities of Minneapolis/St. Paul near the confluence of the Minnesota River and the Mississippi; Cincinnati in the deep heart of the Ohio Valley. There are the cities on the small rivers: Indianapolis, Dayton (Ohio), Fort Wayne (Mich.), Columbus (Ohio).

Chicago, areally Cook County, more commonly the Chicago area, and popularly known as Chicagoland, is one of the unique urban pheno-mena of the New World (Figure 32). As with Los Angeles, it is a com-munity of communities, a loose sprawl of districts lacking integration. But in contrast to Los Angeles, Chicago is a great regional focus, com-manding a tributary area some 500 miles in radius. The setting is a series of low lacustrine terraces along the shallow, open waterfront of Lake Michigan. The core is located around the small Chicago River estuary. Chicago was born here, the controlling railroad terminals have thrust in here, and the port function does not cease to develop here. The urban area is oriented to the waterfront and advances lakewards as well as landwards. Intakes from the foreshore have enabled the construction of an arterial

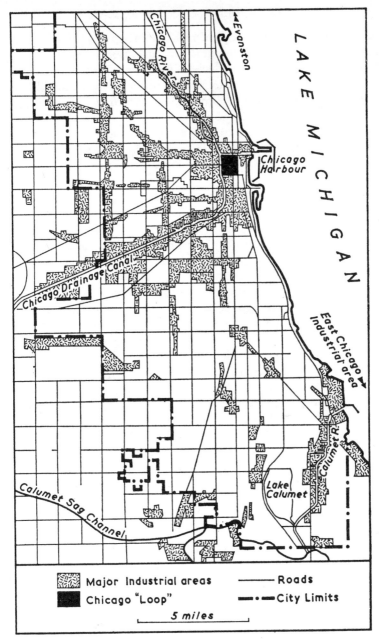

Figure 32 *The Setting of Chicago* (Ill.)

Chicago lies principally between Lake Michigan and the heights of land along the Valparaiso moraine. The urban area has well-defined industrial tracts

highway which binds metropolitan Chicago more closely together. The built-up area extends from the heavy industrial complex of Gary, Hammond, East Chicago, Calumet City (with Lake Calumet offering possibilities for a major port terminal) in the south-east, through a succession of riviera residential districts to the wealthy suburban lakeshore estates north of Evanston. The distance varies according to the terms of measurement, but the urbanized waterfront stretches through at least 30 miles. Downtown Chicago, the Central Business District, burrowing its foundations 80 feet down to bedrock, has experienced that upward translation into skyscrapers which typifies most American cities. There is a fairly distinct break at the 200-foot building level in a city where single dwellings still take precedence over apartment buildings. Peripherally, Chicago is extending itself in new 'surveys' and 'developments' with all the vigour that realtors (or estate agents) can inject into industrialists and hire-purchase home-builders. It is often a development ignorant of those innuendoes of slope which make all the difference between dry sites and wet sites. Lake Michigan guarantees the city an adequate water supply in so far as there is no major pollution. As part protection against this, the Chicago Drainage Canal, following the line of the old Illinois-Michigan Canal (1848), carries the city's effluent over the height of land to the headwaters of the Illinois system. In the built-up area Chicago experiences swift changes in land values according to shifting ethnographic distributions and in few cities has there been such a prodigious number of social investigations.

Changing values naturally affect all forms of settlement. The Middle Western lands have their areas of declining attraction and growing appeal. Most commonly these are related to changing resource values or to the exhaustion of raw materials. Well-defined examples of decline are the lumbering settlements in up-state Michigan and the copper settlements in the Hancock peninsula. The relative status of individual cities is frequently fortuitous, springing out of some local invention or initiative, e.g. Detroit. Fiction has nothing which cannot be matched by fact; but Middle Western settlement has been served well by fiction. Its small-scale features are immortalized in Sinclair Lewis's *Main Street* and Sherwood Anderson's *Winesberg, Ohio,* while the names on the land—Kokomo Kankakee, Oscaloosa and Kalamazoo—are as musical as any in the world.

THE ECONOMY OF THE MIDDLE WEST

The economy of the Middle West is diversified and well balanced. Its primary resource is the land—a storehouse of fertility, maintained with increasing care. Beneath it are varied sources of wealth which provide bases for industry, while a well-integrated transport net enables deficiencies to be easily imported and surpluses to be dispatched. The Middle Western area strikes a tolerable balance between urban and rural residence. The heterogeneity of origin and relatively recent arrival of many urban inhabitants stand in contrast to the comparatively old-established rural settlement. The country-dwellers are mostly Americans of American antecedents.

In Chicago Art Gallery hangs Grant Wood's *American Gothic* (1930), depicting a poker-faced Middle Western farmer and his sturdy bonneted wife, with farm implements to hand. It might well be sub-titled 'Hoe your own row'. For Middle Western farms are commonly owner-operated, and there is a general absence of rural employees (save for the occasional 'hired man' who is probably a younger member of a neighbour's family). Investment per head is very high. Farm buildings must be substantial, for over almost the entire area it is necessary to house animals throughout the winter. The land lends itself readily to mechanization; but equipment can be expensive. The Middle West is one of the homes of the tractor and an inventive hearth of much agricultural equipment from corn-huskers to pea-picking machines. Hire-purchase commitments are increased in the cause of mechanization, though rural processing plants frequently contract for crops in the field and send their own harvesting machines. The individualism of the area may be modified in sales organizations. The dairies of Wisconsin and Minnesota, with their butter- and cheese-making plants, provide examples of a measure of regional co-operation. The roadside booth, with the season's fruits and vegetables, honey, cider and syrup, continues to reflect the attachment to individual marketing, while the rural competitive spirit reaches an August peak in the annual state fairs.

From the outset, Middle Western farming has been mixed farming, with stock and crops common to almost all holdings. Natural grassland was swiftly converted to a rotational system of arable cultivation in which Indian corn took the lead. There can be few agricultural regions which have made such a world impact in the space of a generation as 'the Corn Belt'. Its vocabulary, writ large from Corn Belt Banks to Corn Belt Beer, is folk nomenclature which was elevated to scientific status by O. E. Baker (Figure 12). The concept of the single-crop region as applied to the Middle West in particular has been vigorously challenged by John Weaver, who favours the substitution of the crop-combination region (Figure 33). When corn is 'knee-high on the fourth of July' or when the tanned fields are rustling ripe in the tangy autumn air, the 'Corn Belt' has a visual reality.

Indian corn is a crop which can be used in many ways. It can be harvested as ripe grain (then stored on the cob in bins for animal fodder) or sent to the mill for processing. Corn meal and corn mash may be by-products of a milling process which may extract corn starch, corn syrup and corn sugar. It may be harvested green for ensilage. It may be fed or

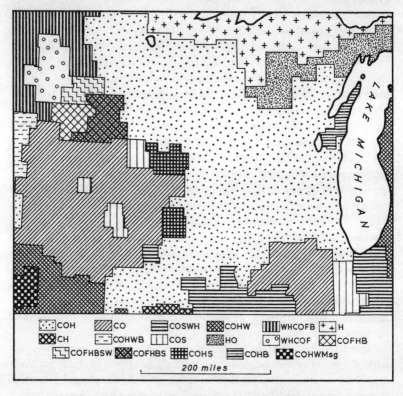

Figure 33 *Crop Combination Regions in the Middle West*

This map reproduces a section of the crop combination map for a part of the Middle West compiled by J. Weaver, 'Crop-Combination Regions in the Middle West', *G.R.* 44, 175–200, Plate 2, A, 1954, reproduced by permission of the author. The key to the dominant combinations is as follows:

C = corn, O = oats, H = hay, S = soy bean, W = wheat, F = flax, B = barley, Sg = sorghum

grazed as a green fodder. The true 'Corn Belt' is spread across the southern tier of Middle Western states, but corn is a common field crop in the northern tier, where Baker identified a 'Hay and Dairying Belt'.

Corn has been varied in the southern Middle West by autumn-sown wheat, barley, oats, legumes and pasture grasses. Since the 1890's, soy beans (*Glycine max*) have been formally admitted to rotations. Soy beans are another multi-purpose crop and for that reason the statistics of their acreages are apt to mislead. They may be harvested as a field crop, milled and used for animal fodder or industrial purposes (e.g. vegetable oil). They may be ensiled or grazed in the field as green fodder. At the same time they are a leguminous plant which makes a positive contribution to the soil. The climate of the Middle West, save in the harsher northern areas, enables not only varied cereal and fodder crops, but also more specialized cash crops to be grown. Some of these have stable markets, e.g. the peppermint and celery around Kalamazoo, the raspberries around suburban Minneapolis; some have a less reliable harvest and market, e.g. the fruits around Benton Harbour.

The predominant cereal and fodder crops are inseparable from the heavy stock, for the southern half of the Middle West is essentially stock-fattening country. An increasing proportion of the stock fattened is bred there, but large numbers still move in from the breeding lands to the west. Corn and pigs still go together, and pigs are to be seen everywhere in the countryside. Pigs also fit well into the dairy farming which distinguishes the northern tier of states. Some Middle West farms are virtually 'livestock factories', sensitive to the marketing figures broadcast daily over the local radio for the local stockyards. Wisconsin and Minnesota are America's leading producers of butter and cheese, and the by-products of their dairies are absorbed in pig production. Cheese marts and creameries along the highways tell of the products of the land, but they do not reflect the tug-of-war between dairying and meat production which takes place on most Middle Western farms. Changing values of milk and meat are sensitively reflected here in changing volumes of output.

The mineral wealth beneath the fertile soils takes varied form and is contrasted between north and south. Clays for brick and tile manufacture, sands for glass, gravels for constructional purposes are generously distributed. Calcareous rocks for conversion into cement and lime occur at varying depths. Limestones and sandstones of varying hardness and virtue are available for quarrying, e.g. the Niagara series in up-state Michigan and the Portland-like stones of southern Indiana. At greater depths are the reserves of the Mid-Continental coalfield. But they are modestly exploited, and their pit headgear sits incongruously in the middle of cornfields. Natural gas occurs widely: oil, intermittently.

To the north, the fertility of the soils diminishes, but the wealth of the rocks increases. The Shield penetrates into Minnesota and Wisconsin to endow the U.S.A. with one of its richest mineral hoards. The quantity and quality of the resource is regionally variable and the scale of operations is adjusted accordingly. Iron ore dominates. There are six ranges, of which Minnesota's Mesabi, Cuyuna and Vermilion are richer than those to the south of Lake Superior (Gogebic, Marquette and Menominee).

The range country of Minnesota, standing back from the rocky waterfront but favoured by a downward slope to its shores, remains America's largest single source of iron ore, despite threat of exhaustion. The best-known of the north-shore ranges is the 80-mile-long Mesabi ridge (Figure 34). It contains some of the largest open-cast workings in the world, e.g. the Mahoning-Rust pit at Hibbing (3 miles long, 1 mile wide and 400-feet deep), though underground mining increases. The Mesabi Range displays workings in all stages of development and decay. Ghost settlements mark exhausted pits; dying towns, such as Cuyuna, with its shuttered shops and weed-infested sidewalks, reflect the ephemeral character of mining enterprise; 'reactivated' sites, e.g. Mahnomen pit, speak of the urgent demand for ore and of new methods of refinement; young towns, e.g. Babbitt, indicate the shift from high ore-content workings (60 per cent or more) to the lower quality ferruginous cherts which are called taconites. Settlements vary in size from such mature cities as Hibbing and Virginia; through established towns which are proud of their mining traditions such as Buhl, 'The Hub of the Range' and Chisholm, 'The Iron Bowl City'; to half-mining, half-farming settlements, e.g. Parkville and parvenu towns like Hoyt Lakes and Taconite Harbour. Coniferous timber, bald rock, muskeg and open lake divide the settled areas. The pits gouge deeply into the face of the land, while their tip heaps and spoil heaps (the product of screening and sintering) contour it anew above the tree-tops. It is a land loud with explosions and the clash of armour-plated equipment, where men go crash-helmeted, where pumps work ceaselessly in rainy times to drain the pools of blood-red water from the open workings, and where in dry times people and places are stained terra cotta with the dust of the oxidized ore. Duluth/Superior, the chief exporting harbour, dispatches about 60,000,000 tons of ore annually during its 6-month period of open water—about ten times the annual export of Narvik.

The Middle West refines the products of its own farm, forest and mine, but it is also concerned with the import and conversion of raw materials.

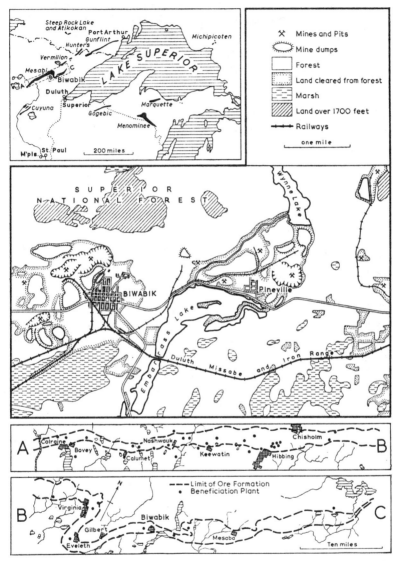

Figure 34 *Biwabik, a Sample Area in the Minnesotan Iron Range Country*

The diagram shows the distribution of the principal ranges, the course of the Mesabi ridge and the detail of land use around Biwabik (Source: *U.S.G.S.*, 1 : 24,000, Biwabik Quadrangle)

Primary and manufacturing industry go hand in hand. Middle Western industry provides examples of old-established concentrations, with heavy investment, which are tied to their original locations, e.g. Chicago's Near South Side or the Ford Plant at Highland Park in Detroit. It provides contrasting illustrations of new industrial areas, in suburban areas and open country which reflect the differential growth of industries not limited in their choice of location by raw materials or by historical facts. The Clearing Industrial District of Chicago is an example, though almost any moderately sized Middle Western city will duplicate it on a smaller scale. Middle Western industries have made a profound impact on the world market and world imagination. The two most typical examples are the agricultural processing and the transport industries. Chicago, with its widespread stockyards, its canneries, tanneries and animal by-products, retains the atmosphere of Upton Sinclair's *The Jungle* and the advantages of scale, though advantages of location are increasingly shifted to cities nearer the concentrations of animals. Detroit persists as the world's automobile capital and it is fitting that Cadillac, who founded the French settlement in 1701, should have his name enshrined in one of America's most cherished cars. Reminders of the motor industry are scattered throughout Detroit, from the downtown General Motors Building, the executive centre of the G. M. Corporation, to the highway rejoicing in the name of 'Automobile Row'. There are, however, three primary concentrations—Hamtranck, with the Dodge Plant; Highland Park with the Chrysler plant; and biggest of all Dearborn, with the Ford River Rouge plant. The River Rouge plant employs, directly or indirectly, about 90,000 people. There are assembly plants at other major cities in the Middle West, e.g. Flint (Mich.) and Minneapolis (Minn.). There are subsidiaries across the Detroit River in Ontario. Inseparable from the automobile industries are the manufacture of aircraft and of agricultural machinery.

CROSSROADS AND CRUX OF THE CONTINENT

The Middle West binds elements of uniformity and diversity together and has maintained a notable stability through successive agricultural and industrial changes. Its continuing equilibrium is closely related to a location in space near enough to serve the consuming market of the North-East with the greatest possible advantage, and a development in time sufficiently delayed to escape the north-eastern problem of obsolescence. The Middle West, adhering more logically to the east than to the west, might not inappropriately be called the Middle East. Certainly the world has few more fertile or energetic crescents of country. The Middle West is the most crucial part of the U.S.A.: it is the crossroads and crux of the continent.

SELECTED READING

A general regional introduction is provided in

J. H. Garland, *The North American Mid-West* (New York, 1955)

Contrasts in settlement patterns are seen in

A. H. Meyer, 'The Kankakee Marsh of Northern Indiana and Illinois', *Papers of the Michigan Academy of Science, Arts and Letters*, 21 (1935), 366–95

H. B. Johnson, 'Location of German Immigrants in the Middle West', *A.A.A.G.* 51 (1951), 1–41

J. E. Brush, 'The Hierarchy of Central Places in S.W. Wisconsin', *G.R.* 43 (1953), 380–402

A. K. Philbrick, 'Principles of Areal Functional Organization in Regional Human Geography', *E.G.* 33 (1957), 306–36

D. R. McManis, *The initial evaluation and utilisation of the Illinois Prairies, 1815–40* (Chicago, 1964)

Contrasts in soil studies are provided by

L. Hewes, 'The Northern Wet Prairie of the U.S.', *A.A.A.G.* (1950), 40–57

H. H. Bennett, *Soil Conservation* (New York, 1939)

Changing methods of studying crop distributions are offered by

O. E. Baker, 'The Corn Belt', *E.G.* (1927), 447–66

J. C. Weaver, 'Changing Patterns of Cropland Use in the Middle West', *G.R.* 44 (1954), 175–200

A new crop is treated in

A. A. Munn, 'Production and Utilization of Soy beans in the U.S.', *E.G.* (1950), 223–34

and a new product in

C. F. Kohn and R. E. Specht, 'The Mining of Taconite, Lake Superior Iron Mining District', *G.R.* 48 (1958), 528–39

See also

E. J. Taaffe, 'Air Transportation and U.S. Urban Distribution', *G.R.* 46 (1956), 219–38

H. M. Mayer, *The Port of Chicago and the St. Lawrence Waterway* (Chicago, 1957)

6

The Old South and the Legacy of the Past

*The Character of the Old South—The People of the South—
Tradition and Change in Agriculture—Tradition and Change in
Industry—'The Land of Flowers'—Organization in the South*

THE CHARACTER OF THE OLD SOUTH

The Old South is another region that takes shape in the minds of Americans, but it is difficult to outline it precisely upon a map. It has been variously defined as the group of states south of the Mason-Dixon line, the eleven states that wanted to secede from the Union (but they included Texas and Texan motives were not identical with those of other states), the historical area of the cotton culture (much changed in its contemporary distribution) and the area in which segregation is practised. The concept is tied to the south-eastern group of states and is inseparable from their humid, wooded countryside. By tradition, the south-east is divided into a northern tier of states (the 'Hill-land South'), which consists of Tennessee and North Carolina, with marginal Virginia, West Virginia and Kentucky: and a southern tier (the 'Plainsland South' or the 'Deep South')—South Carolina, Georgia, Alabama, Mississippi, Louisiana, with marginal Arkansas and Florida.

The differences between the development of the South and of the North-East or Middle West, are inseparable from the nature of its climate and the course of its history. The legacy of the past is reflected in the divided society and the time lag in economic evolution. The climate is productive, though in the very fecundity which favours plant growth there is something that engenders human lethargy. Such an influence cannot be measured scientifically, but it diffuses a leisureliness which contrasts strongly with the bustle of the north.

The map of the south-east is physically dominated by two features—
the Appalachians and the Mississippi Valley. The Appalachians reach their

greatest breadth and highest altitudes in North Carolina and Tennessee. The crystalline rocks of the Blue Ridge rise to monadnock peaks of 6,000–7,000 feet in a zone 70-miles broad, and the Great Smokies of Tennessee provide turbulent scenery of equal impressiveness. Westwards, the corrugations of the barren sandstone ridges and fertile, weathered limestone valleys yield to the smoother sedimentaries of the Appalachian plateau (altitudinally about 2,000 feet) in Tennessee and Kentucky. Still farther west a wealth of limestone features is displayed in the interior plateaux. Mammoth Cave Quadrangle of the U.S.G.S. topographical survey claims 2,800 sink-holes as well as the great caverns from which it derives its name. Karstic phenomena of Kentucky are seen from the air in Plate 4.

The arc of five states from Alabama in the south-west to Virginia in the north-east share in the Piedmont, with its adjacent Fall Zone, 150–200-mile-wide coastal plains and swampy, islanded coast (Figure 35). The hilly Piedmont, with sand and clay loam soils, gives way to the more steeply pitched Fall Zone peneplain into which the succession of sizeable rivers have eroded cataracts and rapids. Rivers such as the Chattahoochee and Savannah also serve as state boundaries. Lagoons and swamps, threaded by the Intra-Coastal Waterway, characterize Gulf and Atlantic shores alike. The 'sea banks' of North Carolina encompassing 25-mile-broad Pamlico Sound and forming Cape Hatteras, with their wooded 'hammocks', swale hay, eel grass, 'gurgitating sands', isolated communities and legends of wreckers, are the extreme expressions of coastal characteristics.

Louisiana, Mississippi, Arkansas and the southern third of Tennessee with its regional capital of Memphis look to the Mississippi. Below the confluence with the Ohio at Cairo, the triangle of lowland that broadens seaward has been moulded in both the physical and human sense by the behaviour of the river. Relics of former river channels, cut-off meanders and ox-bow lakes proliferate. They are graphically inscribed in the scrolls and swirls of contours, for example, around Clarkesdale (Figure 36). Tributaries flow in principally from the west—the lesser White River from the Ozark plateau, the Arkansas which breaks through the Boston and Ouachita mountains, the Red River which divides Louisiana. The east bank receives, among others, the Yazoo, the valley-bottom lands of which are among the most agriculturally esteemed in Mississippi. The historic western boundary of the Old South is the Sabine River.

The Gulf coast states display a surfeit of water. They embrace some of the south's largest lakes, e.g. Grand Lake in Louisiana and Lake Pontchartrain to the east of New Orleans. Settlement is very sensitive to the water-

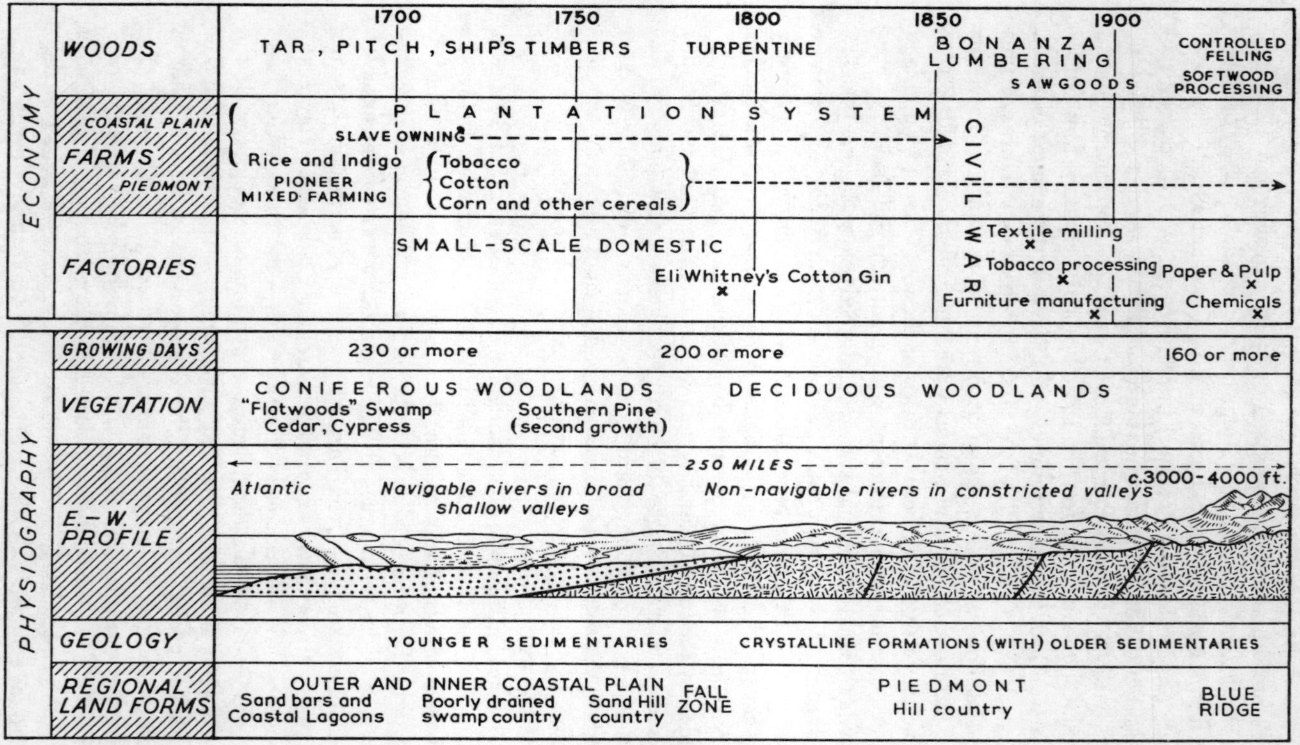

Figure 35 A Transect through North Carolina

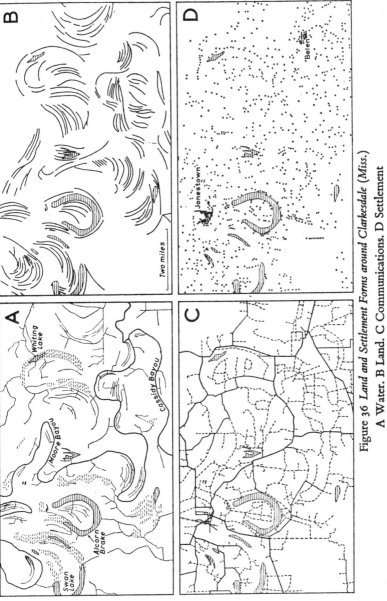

Figure 36 *Land and Settlement Forms around Clarkesdale* (Miss.)

A Water. B Land. C Communications. D Settlement

The flood plain, scrolls and swirls of the Mississippi and its tributaries are illustrated here, side by side with the human response to their forms. The low, swampy areas assume the *bayou* name (Source: *U.S.G.S.,* 1 : 62,500, Marks Quadrangle)

ways and the minor relief features lifted above them. The *bayous* have provided linking routes for those familiar with their watery mazes as the rivers have offered the historical (and, in contrast to the Atlantic seaboard, navigable) channels of entry. The successive *cheniers*, or raised beaches inland from the Louisiana shore, are picked out as the oak ridges in the swampland vegetation. Evergreen or deciduous oak, hickory and pecan are festooned with Spanish moss no less than the swollen-trunked cypresses. The *cheniers* support strings of settlement; while around Baton Rouge, the dissected edge of the youngest Pleistocene terrace has a scarcity value for 'country seat' properties. Roads may strive to follow these features, but are eventually forced to use elaborate viaducts and bridges. Lake Pontchartrain has a 20-mile bridge. The four-lane highways are alien to the steamy and primeval natural landscape which with its palmetto and sawgrass, water lilies and white herons, squat pelicans and occasional alligators, cannot be very different from when colonists first saw it. The Atlantic coastal plain of Georgia and the Carolinas repeats the swamp formations, from the Okefenokee Swamp in the south to the Dismal Swamp on the Virginian border.

The South rises to climaxes in Appalachia or along the Mississippi. In 24 hours a visitor may journey to half a dozen ante-bellum mansions, where classical porticoes retreat behind magnolias, rhododendrons and live oaks. He may trace in farm patterns the influence of the French *arpentes* (Plate VI) and still hear a French *patois* spoken in Louisiana. In Appalachia the split-log cabin, the tortuous snake fence, the 'moonshine' manufactury (or distillery) and the horse-and-buggy may still be encountered. But it is traverses between such cities as Mobile and Birmingham, Savannah and Atlanta, Wilmington and Durham that reveal the more representative background to daily life. The land escapes almost imperceptibly from ill-drained through better drained to the long, loping hills that anticipate the Piedmont. Under the plough, it is swift to seam with gullies, and assumes the bright oxidized red associated with lateritic soils. The line of the woods is everywhere on the horizon, but because most cleared land lies along the ofte·· unsurfaced roads, the fact that two thirds of the area is forested is only apparent from the air (Figure 37). Small holdings outnumber large holdings, old settlements are generally more important than new, small towns are rarely overshadowed by big cities. Markets which emerged in horse-and-wagon days lie at a comfortable 25 miles from each other. Roadside booths sell farm produce, spring-time strawberries, summer melons and peaches, autumn pecans. Shirt-sleeved (and 'shirt-tail') people sit a good deal on verandahs in a

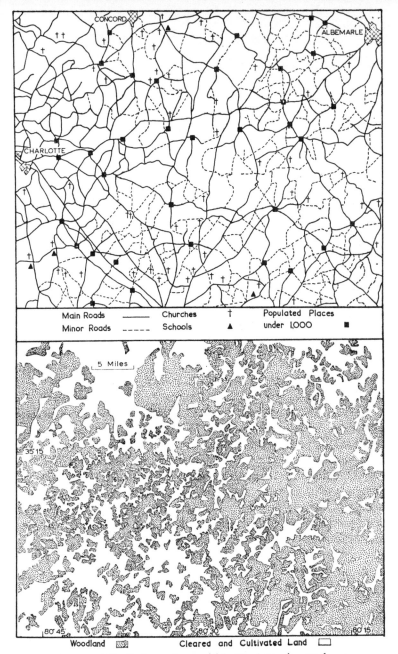

Figure 37 *The Settlement Pattern of the North Carolina Piedmont*

The even spacing of settlement nodes and the complementary pattern of woodland and of cleared land reflect essential details of the Carolina piedmont. The mesh of communications predates the establishment of township and range lines. S. T. Emory, 'Topography and Towns of the Carolina Piedmont', *E.G.* 12, 1936, 91–7, analyses certain of these distributional features (Source: *U.S.G.S.*, 1:250,000, Charlotte Quadrangle)

greenhouse atmosphere. Their diet, heavy with the starch of cornbread, 'grits', sweet potatoes and black-eyed beans, is relieved by half a dozen different kinds of kale and gourds. For the medical geographer, their fevers and parasites trace new patterns to correlate with the regional inertia and with the problems of William Faulkner's Yoknatapawpha country.

THE PEOPLE OF THE SOUTH

The south-eastern group of states is inhabited by a fifth of America's population. If the main criterion of being American is to be born in the U.S. of U.S. parents, this is one of the most American parts of America. In part it is explained by the fact that the Old South was by-passed by the mainstream of late nineteenth-century immigration. For many reasons, it remains an area of emigration, although Florida stands apart from this generalization. The birth rate is high and in some parts as high as anywhere on the continent: the death rate is relatively high. A full third of the inhabitants are coloured.

America's ethnographic dilemma is greatest in the south-east, the home area of two thirds of its Negroes (Figure 38). There are about as many coloured people in the U.S.A. as there are Canadians in Canada. Most of the coloured people in the South are the descendants of the 4,000,000 slaves who were freed in the 1860's. The concept of 'colour' is defined in the preamble to the volumes of the Census. The South has received a limited immigration of mixed people from the Caribbean and, by definition, the Creoles of the Mississippi delta form a distinctive group. Considerable admixture has given rise to a large mulatto element, who are grouped with the Negroes. Within the coloured community, there is a very finely differentiated assessment according to such characteristics as pigmentation and hair form. The percentage of coloured people in different states varies substantially. In Mississippi and South Carolina, it may exceed 40 per cent; while in Virginia it falls to 27 per cent, and Tennessee, 18 per cent.

In the southern states population shows a fairly even distribution between rural and urban residences. The distribution of the coloured population is rapidly shifting from rural to urban areas. In 1910 one quarter of the Negroes lived in urban areas; today, more than two thirds live in towns. The dichotomy between white and coloured settlement is more pronounced in towns than in rural areas. Distinct Negro districts evolve, and where a small town becomes an expanding local shopping centre, e.g. Selma (Ala.), it is possible to see the emergence of white and coloured

shopping streets in process. Segregation is the order of the day, from bus
and train to restroom and restaurant. In rural tracts, shanty-town hamlets
spring up, e.g. in the tobacco land of the Carolinas, which may be bereft
of essential services for the living and of suitable graveyards for the dead.
Elsewhere, the derelict clapboard cabin of the Negro may sit within

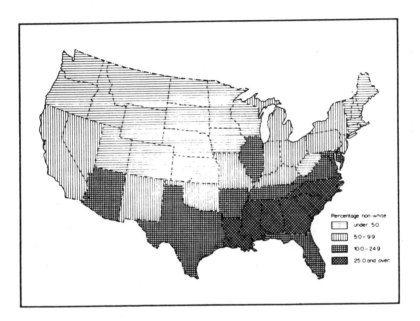

Figure 38 *The Percentage of Non-White Population in the U.S.A.*

The distribution of the non-white population of the U.S.A. has been profoundly
modified during the post-war period. The number of Negroes outside the
states of the Old South now exceeds the number inside those states (Source:
Hart 1960 and U.S. Census of Population 1960)

sight of the well-maintained white farmhouse or, when the Negro
migrates, may be diligently removed by the state. For over the last generation the coloured southerner has been increasingly on the move—first
from the hills, next from the 'Deep South' and the central south, now the
trans-Mississippian south. Movement is essentially to urban areas outside
the South. The third element in the social structure of the South is, in some
respects, the most difficult. It is the 'poor white'—who arrogates to himself

a superior status to that of the Negro, but who is socially rejected by the rest of the white community. It is among the 'poor whites'—who may be known locally by a variety of names varying from 'hill billy' to 'clay eater'—that coloured antipathy reaches its strongest expression.

The Old South lacks large cities and conurbations. The largest is New Orleans (pop. *c.* 750,000). It is within the meanders and below the level of the Mississippi River; the lowest bridge point; an ocean-going port; its old quarter architecturally reflective of Spanish and French antecedents, and one of the two most distinctive cities in the Union. It has a very heterogeneous population, a legacy from its past as well as a result of its present situation upon the threshold of the Caribbean. Yet New Orleans is neither central to the south-east nor typical of it. In size, the only rival is Birmingham (Ala.) which might lay claim to the status of regional capital if it had something of the historial background of a city such as Atlanta (Ga.). For the rest, urban centres may be state capitals, e.g. Jackson (Miss.), Montgomery (Ala.), strategically located industrial towns such as the 'Fall Zone' cities of Macon or Augusta (Ga.), Columbia or Raleigh (N.C.); or historic ports the mid-twentieth-century development of which never quite matches up to their earlier paramountcy. Savannah, once America's chief cotton-exporting harbour, and Charleston, apotheosis of the 'colonial' South, provide examples.

TRADITION AND CHANGE IN AGRICULTURE

The Old South is essentially agricultural. It depends more heavily upon farming for its income than any other region. The characteristic is evident in the numbers of rural versus urban residents, rural farm versus rural non-farm inhabitants, and in a landscape where country takes absolute precedence over town. Agriculture is conducted in a heavily wooded countryside, where soils are generally poor, drainage frequently bad but where the climate is kindly.

In the ante-bellum period the South was called plantation or planter-slave country, but even before the break-up of the old system it supported large numbers of smaller holdings. Widespread tenancy and share-cropping succeeded to the plantation system. The splintering of much of the land into tenant holdings, most of them without capital to invest in seed, fertilizers or essential equipment, had a serious effect upon the rural economic structure. As recently as a generation ago, for example, Mississippi listed 72 per cent tenant farmers, with 60 per cent of its tenants paying in kind. Share-cropping with the landowner providing land, seed, animals, fodder, housing, equipment and fertilizer and receiving a half share of the crop is a continuing feature; but this must not be exaggerated. Tenancy remains one of the least satisfactory features of farming in the south-east. It is usually short-period tenure and is accompanied by frequent changes of operator. When a farmer exhausts his tenanted land, he moves on. Long-term projects and experiments are inhibited accordingly.

The South has been a land of staple crops—corn, cotton, tobacco, sugar and rice. Staples persist, but they no longer dominate particular areas as much as formerly; among them corn and cotton lead. Cotton has been the chief cash crop; corn, the chief food crop. Cotton is an annual plant and is harvested between August and October. A century ago, a kind of shifting cultivation was commonly practised, with the cotton fields rotating around the area of the estate. Today, cotton is both intertilled and grown in rotation. It is relatively heavy in its demands of labour, though most operations are capable of mechanization. In the humid south-east, slope, soil, rainfall conditions and size of farm units all restrict the effective application of cotton-picking machinery. By contrast, picking machines are well adapted to the extensive irrigated cotton lands of Texas. The traditional pickers have been the Negro tenant farmer with his family, but many of these are leaving the land for other pursuits while the more enterprising are changing their methods of farming. Town-dwelling cotton pickers are trucked daily to and from the fields. For example, as many as 20,000 pickers a day may move out of Memphis (Tenn.) at the height of the season. 'Day haul' workers may also include seasonal immigrants from Mexico or the West Indies. In Arkansas, for example, as many as 23,000 Mexican cotton pickers may move in for the cotton-picking season.

America still leads the world as a cotton producer and exporter. The south-east still supports extensive cotton lands, and 'cotton farms' (drawing more than half their income from the crop) are common. But there has been a sharp contraction in cotton acreage since the 1890's. The contraction is inseparable from the invasion of the boll-weevil which thrives in the warmer, moister parts. Direct assault on the pest by poison spray, by inploughing, or by firing the autumn stubble have been less effective than the introduction of new crops to rest and clean the land, and of new quick-ripening cotton plants which mature more speedily than the weevil. Although plenty of cotton is still grown within the confines of the historic 'Cotton Belt', e.g. inner coastal plain of Georgia, specialized cotton growing has migrated to the drier west, the cooler Piedmont and

Tennessee. The control of King Cotton over the southern farmsteads has been relaxed. Enterprise, a southern Alabaman town, even has a monument to the boll weevil because it eliminated the hazard of cotton failure and forced the community to turn to the more reliable and remunerative peanut (*Arachis hypogaea*).

Tobacco is a second staple which stamps the southern scene. Tobacco is an annual crop, commonly harvested in August, and it derives from farms of many different sizes and ownership conditions. There are distinct regional specialities, from burley tobacco (for chewing and pipes) in Kentucky, through bright-leaf tobacco (for filling) in Virginia, to flue-cured tobacco (for high-quality cigarettes) in North Carolina. Pickers and tiers are drawn from both local sources and the army of migratory field workers. Drying, flue-curing and 'maturing' sheds sit in the middle of Piedmont tobacco fields as well as in and around the many small tobacco-marketing towns which thrive upon the crop. Across the range, Lexington (pop. 54,000), in the Kentucky Blue Grass country, claims the largest market for loose-leaf tobacco in the country.

Sugar is a third southern staple, but its cultivation is restricted by climate to the Gulf margins. Wild cane is a component of the 'good woods' of the lower Mississippi Valley and graziers have long regarded the native 'cane brakes' as the best natural cattle fodder of the area. Cane cultivation is usually a large-scale operation, incorporating a refining plant. The National Sugar Refinery at Godchaux, north of New Orleans, provides an example. Extraction of syrup is also undertaken by small operators.

Traditional crops, property forms and methods of cultivation persist, but they are intruded upon by new. Diversification and mechanization lie behind the processes of change. The one crop economy which dominated much of the historic south not only enforced a degree of dependence upon the market for cotton, but was also responsible for the serious exhaustion of soils which were often far from fertile in their original form. Undoubtedly the methods of cotton cultivation caused much of the erosion which has converted formerly clear rivers into muddy waterways. No region of the U.S.A. has a greater proportion of land subject to erosion than the south-east. Removal of the protective timber and exposure of soil are primary causes. Diversification can provide a cover crop in rotation with cotton or a subordinate crop in association with cotton. In any case the length of the growing season enables a second crop after the cotton harvest in many parts of the south. The principal crops which have been introduced are improved cereals, legumes and improved grasses (e.g. Bermuda grass, Dallis grass, Johnson grass).

Much land has been taken out of row crops and sown to improved grass in the last generation. Together with new crops this makes stock-keeping increasingly practicable. There has been no widespread tradition of animal husbandry in the south-east, and most meat and dairy products were formerly imported to it. When cotton prices were high and before the advent of the weevil, animal husbandry was scarcely competitive with cotton cultivation. The humid climate favours a swift growth of fodder plants for controlled grazing and ensiling. Combinations such as green corn and velvet bean may be used in summer and provide an almost jungle-like growth: while in winter, small grains may be grown as a cover crop for feed. Animal fodders also derive directly and as a by-product from peanuts, soy beans and their vegetable-oil industries. Mild winters reduce stabling needs to a minimum. Dairy cows are not entirely suited to sub-tropical summers, but beef cattle (Hereford, Angus, Brahman and crosses) are the subject of continuous experiment among the new settle-ments of the South. Pigs are the most satisfactorily reared of all southern farm animals, while few parts of the U.S.A. have developed the broiler industry so intensively as the Piedmont. The last generation has witnessed an all-round growth in numbers of farm stock, except for horses and mules. Animal products now account for almost half the farm income of the south-eastern states.

Escape from the shackles of a one-crop system also reduces depend-ence upon artificial fertilizers. No region has applied artificial fertilizers to its soils more heavily than the south-east. Clovers, beans, alfalfa (intro-duced 1730's), lespedeza, soy beans and peanuts (introduced 1915) have been important soil restoratives.

Crop diversification and stock multiplication have spread demand for labour more evenly over the year. Hitherto, over-employment during the period of tillage and cotton picking was followed by under-employment during a 5- or 6-month 'dead' season. The new stock and crop relation calls for skilled and flexible farm labour, but guarantees regular employ-ment. When the Negro small holder is idle from November to March, the livestock farmer is busy with his green rye grazings, Bermuda grass and lupins. The need to learn new skills is also inseparable from the intensi-fied mechanization of farming in the south-east. Mechanization offers new openings for a limited number of skilled workers, but closes the door on many others and causes redundancy.

Mechanization takes two principal forms. There is first the employ-ment of mechanical equipment such as tractors and steam pumps for general farm purposes. In the south-eastern states in 1950 only one farm in seven had a tractor compared with one in two for the nation at large. The neglect is being remedied. There is secondly the provision of specialized machinery for cash crops. Application to the two principal staples, cotton and tobacco, has been slow. Tobacco is particularly difficult to harvest by machine. Cotton-picking machines represent a very heavy capital outlay and may still have to be supplemented by hand labour, e.g. on wet soils or if high winds disturb the crop. In any case, it is better to use hand picking to remove the first bolls. One machine can do the work of forty to eighty pickers. Rice planting and harvesting has been satisfactorily mechanized in the 'Deep South'. Soy bean combines, peanut shakers, corn harvesters and grass planters are successfully employed. Mechanization has at the same time released much land hitherto grazed by draught animals. Land formerly used to feed mules may now feed cattle.

Southern farming is in transition. The means for its transformation are available: the ends await fulfilment. As yet the potentialities of the South are rarely achieved except on the neo-plantations and larger owner-operated farms. Wealthier agriculturalists, hobby farmers and industrial organizations employing manager operators have both compacted together large-scale holdings and fitted new enterprises into the frames of ante-bellum estates. All the opportunities offered by diversification and mechanization are being explored on holdings that may be 1,000 acres or more in size. In some counties the new agrarian aristocrats may control a third of the surface area. The plantation *redivivus* has another effect. Its practices are the object of emulation by middle-sized farm operators. The diffusion of its ideas will help to lift their yields of hay above a ton to the acre and their milk output by a fifth to that of the national average. By contrast it has relatively little to offer to the 50-acre Negro tenant farmer who still moves at the pace of the mule-drawn mouldboard plough.

TRADITION AND CHANGE IN INDUSTRY

The South lacks a strong tradition in manufacturing industry. There is no single explanation. Historically there was little incentive, for it had important staples to export in exchange for essential needs and luxury manufactures. Secondly, it was deficient in the widely distributed industrial fuels available in the north. Thirdly, the Civil War intervened at the most critical phase of its industrial evolution. The financially exhausted south had no funds to invest in industry, and the victorious north was not interested in creating rival factories in it. Fourthly, the immigrant factory

worker from Europe was more attracted to the North, where the choice
of employment was more varied. The growth of a tradition in the North
cast its shadow over the South, where technical knowledge lagged behind.
As expressed in the number of patents granted, inventive skill remains at
a premium in the South. Moreover, the industrial tradition of the South
today is essentially a white tradition and the percentage of coloured factory
workers in any of the better-paid positions is very small. Climate may also
have discouraged large-scale factory industry, but if so that of the semi-
arid south-west has done nothing to inhibit it.

Some industries are indigenous to the South, some have migrated to
it, some have been created in it. Tobacco manufacturing, seed crushing,
vegetable-oil extraction, lumbering, wood-working, cotton ginning and
metallurgy are indigenous. Abundant and varied timber and a wood-
working tradition have aided the rise of furniture making. By the end of
the eighteenth century, Thomas Asche's *Narratives of Early Carolina* (1650–
1708) tell of 'tables, scrittors and cabinets' made from 'sweet-smelling
cedars'; the Carolinas today claim first place in the industry. Hickory
(N.C.), which once specialized in hickory wagons, is a good example of
continuity in wood-working. Most producing centres are on the Piedmont, e.g. High Point, Thomasville, Mebane, but some are on the coastal
plain, e.g. Dunn. Tobacco processing reaches a peak of importance in
Winston-Salem (N.C., pop. 86,000) and Durham (N.C., pop. 70,000).
The iron and steel industries, around Birmingham, Bessemer and Leeds,
employ local ores and coals and have the strongest raw-material location
of any North American steel complex. They account for about 7 per cent
of America's steel output and specialize in pipe and tube production. Their
setting is disadvantageous for the international market, and beneficiation
costs are high, but they are especially important for the development of
the south-east.

The first significant large-scale migration has been that of the textile
industries, to the Piedmont mills of Virginia, the Carolinas and Georgia.
America's greatest output of cotton textiles today derives from the
historic cotton-producing states: textile manufacturing claims most of
their industrial employees. The factories are larger and more efficient than
those in the North, but they are less concerned with quality products.
Woollen manufactures for the local market have grown apace. Rayon
fibres are produced at Erika (North Carolina). New synthetic fibres derive
from an orlon plant at Camden (South Carolina) and from an acetate
plant at Rock Hill (South Carolina). Ready-made clothing, knitted goods
and hosiery are products of smaller as well as larger towns. For example,

textile mills by the score may be counted in and between almost every settlement from Durham to Marion (N.C.).

Diversification has been strengthened by the introduction of new industries. In part, these have been the offspring of the Tennessee Valley Authority: in part, they are a legacy of the war. Cheap hydro-electric power in Tennessee, North Carolina and Alabama encouraged industries in need of abundant energy, while isolation favoured the location of armament industries. There are nuclear-fission plants at Aiken (S.C.) and Oak Ridge (Tenn.), an aluminium reducing plant at Alcoa (Tenn.), fertilizers and munition in combination at Muscle Shoals (Ala.); while immense chemical plants line the Kanawha valley of West Virginia. Old folk ways and new industrial processes mingle in settlements such as Canton (N.C.) or Calhoun (Tenn.), which the Bowater concern has rescued from oblivion with the largest single softwood pulp mill in the U.S.A. In the paper mill at Acme on Cape Fear River, pines are basic to a very different industry from that which gave North Carolina its nick-name, 'the Tar Heel State'. Specialized refineries are related to local mineral resources, e.g. petroleum at Baton Rouge (La.), sulphur at Grande Escaille (La.) and bauxite for aluminium at Grammercy (La.). Branch plants make an increasing appearance, e.g. agricultural machinery in Memphis (Tenn.), tyre manufacture at Tuscaloosa (Ala.). The Blue Grass Basin of Kentucky is second only to Scotland as a producer of whiskey (bourbon), while Atlanta (Ga.) derives wealth as the home of Coca Cola.

Industry in the Old South has not made and still does not make the ubiquity of impact that it makes in the North-East. It is not lacking in examples of advanced technology, but they are not widely distributed. There may be skilled labour, but compared with the total labour supply, it represents a low proportion. In such a state as Georgia there are pro-bably twice as many young people growing up on farms as there are farm jobs. They may constitute a potential labour pool, but they are likely to suffer under-employment unless they are specially trained and unless abundant capital is made available to equip them. Industry established in the South is up-to-date in equipment, flexible in structure, and in a stronger position to withstand depression than older-established industry in other parts of the country. In theory, there is nothing to prevent the South-East from achieving that balance in agriculture and industry which is the basis of Middle West prosperity.

It is debatable whether Florida, southernmost state of the U.S.A., can be considered a part of 'The South'. It is in it, but not of it. Mainland Florida is the natural prolongation of Georgia and Alabama, but for most people the state is usually associated with the 400-mile-long peninsula, which is separated from the settled areas of Georgia by 'dismal' swamps. The low-lying limestone peninsula is rippled along its margins with raised beaches and sand bars, and exceeds 250 feet a.s.l. only in its central part (Figure 39). Well-drained and ill-drained lands are juxtaposed. Rounded limestone hillocks divide solution lakes from swampy hollows in which lilac-coloured water hyacinths proliferate, ephemeral lakes where the ibis wades, and 'prairies' the moist grasses of which represent the next stage in the botanical succession. Drainage difficulties partly reflect slight gradients; partly, prevailing humidity. For Florida lies upon the threshold of the humid tropics. Caribbean plants in the natural vegetation of the Everglades or the hibiscus in the front garden are good climatic indicators. Precipitation is about 50-60 inches p.a., falls chiefly between June and September, and as much as a fifth of it may fall in 24 hours. In addition to hurricanes, Florida also suffers from the threat of frosts. Even the lower half of the peninsula escapes only about one year in four.

Florida's detachment and its difference from the historic South both account for its delayed development. Ninety years ago most of it lay beyond the frontier of settlement and Seminole Indians still constituted a hazard. Despite the scenic illusion conveyed by the festoons of Spanish moss and the presence of Stephen Foster's immortal Suwannee River, Florida shares little of the culture of the 'Old South'. There is no legacy of the Civil War, while the Negro is immigrant rather than indigenous. Florida is, in fact, an appendage of the north. Steep winter temperature gradients between Florida and the 49th parallel favour this role. Railroads carry its winter products to northern markets and bring northern visitors to its warm shores. It is not much more than a thousand miles by air from either New York or Chicago. The established seasonal migration is accompanied by a substantial permanent immigration, for Florida offers financial attractions (through favourable state taxation) as well as climatic appeal. With 5,000,000 inhabitants, it is among the more populous states of the union and its rate of increase 1950-60 was the largest. Because of the nature of the immigration, it is also among the wealthiest.

Settlement is predominantly urban, strongly concentrated in the peninsular counties and primarily coastal. A succession of major resorts extends south of Jacksonville, through historic St. Augustine, Daytona Beach, Palm Beach, Fort Lauderdale and Miami to the Florida Keys,

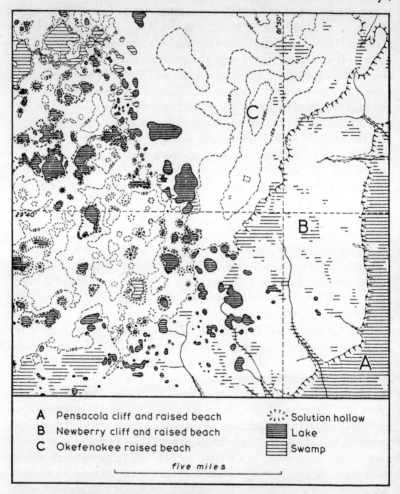

A   Pensacola cliff and raised beach
B   Newberry cliff and raised beach
C   Okefenokee raised beach

Solution hollow
Lake
Swamp

five miles

Figure 39 *The Topographical Features of Lakeland (Fla.)*

Three distinctive landscapes are evident in this Floridan tract—the eastern swamplands, the central sand ridges of former marine levels and the complex pattern of solution lakes to the west. The woodlands of the sand and swamp country are complemented by the orchards and grazing country of the limestone spur (Source: *U.S.G.S.*, 1 : 62,500 Interlachen Quadrangle)

linked by a motorway to the naval base of Key West. These resorts, with their offshore fisheries (from sharks to shrimps), snook and tarpon, luxury hotels and motels, cruisers and yachts are served by a protected intra-coastal canal. The Gulf coast lacks the continuous band of settlement, but there is a concentration around Tampa Bay in Tampa and St. Petersburg. Inland towns are associated with administration and education, e.g. Tallahassee and Gainesville, or agriculture, e.g. Orlando. Few parts of America have experienced such a succession of booms in land values. Most urban transformation has taken place in the last 50 years and has been only intermittently halted by the havoc of flood and hurricane. Contourless development sites, e.g. Boca Crega Bay, near St. Petersburg, are opened up by miles of man-made canal side by side with macadamized roads, for access to water has an amenity value almost as high as access to a highway. Sawgrass and palmetto plam give way to gardens of sub-tropical exuberance. On the cheaper sites, the wise build their homes upon stilts or piles. Yet the sun rules, and a day on which it does not shine is rare. As with the French Riviera the tourist season has been much extended in recent years.

Climate, indeed, is sometimes described as Florida's chief resource. There are few minerals of consequence, and no sources of thermal energy. From deposits around Tampa Bay about three quarters of America's raw phosphates are drawn. Limestones are quarried for cement and construction work and shell banks are heavily exploited as around other parts of the Gulf coast. Many would regard the climate as enervating rather than energizing, but it has compensations, e.g. a year-long growing season. Soils raise problems. Locally, they may be good: those derived from the weathered Hawthorne formations of the peninsular backbone, for example. Water surfeit is the main difficulty, for the greater part of the state has wet humus soils. Individual enterprise and local reclamation projects have been superseded by the larger scale labour of the U.S. Corps of Engineers and federal sponsorship of the Central and Southern Florida Project. Lake Okeechobee is central to a thousand square miles of peat and muckland which it is the concern of the central government to drain and conserve. Organic soils shrink and their cover rapidly dwindles unless they are carefully farmed. Fire may also damage them. This Everglade region is unique and coveted by naturalists for alligators, ibis, spoonbill and egret.

The crop for which Florida is most famed is restricted to a limited part of the peninsula. Formerly the cultivation of oranges extended farther north, but the Great Freeze of 1894-5 desolated the fruitland and new groves migrated southwards. The first groves are encountered on locally favoured hills near McIntosh, but the Citrus Tower on the Floridan ridge near Orlando marks the heart of orangeland. Dark green groves, interspersed with blue lakes edged by white limestone shores, extend to the horizon in all directions. It is the world's most extensive citrus tract and from it comes a third of the world's orange crop. Trees stand about seventy-five to the acre and are clean cultivated. Picking continues from November to April, with December and January as the peak months. Much of the harvest is contract picked, with conveyor belts loading the crop on to 25-foot lorries. A good deal of the crop is processed for carotin, dyes, vegetable fats, alcohol, yeast and molasses, in addition to orange juice. Orange-blossom honey is also much esteemed. Burners strive to keep occasional frosts at bay and a difference of a few tens of feet in altitude may be sufficient to spell failure or success. Orange trees can stand up to 5 degrees of frost for several hours, but more will leave them brown and shrivelled.

As in the Old South, farm practices are becoming more diversified. Cattle are now widespread. Angus, Hereford, Shorthorn, Brahman and Crossbred graze throughout the year on the sown grasses of large-scale holdings. Estates such as the Anthony Farms of the northern limestone ridge name Pensacola-Bahia as the best carpet grass and 2 acres will support a beast throughout the year: in the Everglades, South African Pangola grass supports one beef animal per acre. Diversification also results in new crops on old land, e.g. Chinese Tung Oil trees (for lacquer) north of Gainesville, or the tree fruit papaya farther south. Old crops have also been established on new land, e.g. celery on the reclaimed Everglades where mobile assembly-line machines prepare the crop for the pre-cooling plant beside the trenches. Sugarcane is among the older-established Everglade crops and, in contrast to truck crops, is usually produced by large landowners such as the U.S. Sugar Corporation. The Corporation undertakes the entire operation from sowing to refining, producing about a quarter of America's cane sugar, albeit with an elaborate system of price controls and subsidies. Its Sugarland Ranch, also embracing extensive pasture-lands, is what a latter-day William Bartram would call an 'Elysium of the Savannah'.

The greater part of Florida is covered by trees, and maps still identify by name the lumber lines penetrating the interior woodlands. Today the hardwoods of the interior are relatively less valuable than the coastal softwoods. Paper and pulp companies, providing one of the few sizable manufacturing industries, have acquired as much as two thirds of the woodlands in some counties, and Florida ranks among the chief producing states after Washington.

American Florida is a logical projection of Spanish Florida. Spaniards appreciated the position of this land of flowers in Caribbean strategy, strove to establish their tenuous missions and modest savannah ranches upon it and seriously sought for the mythical fountains of eternal youth within it. Americans through agricultural and horticultural endeavour have given heightened meaning to the word Florida, have lifted to new levels the cattle industry, have established playgrounds where they strive to keep old age at bay, and have created a Cape Canaveral from which they may be destined to play Columbus in new spheres.

ORGANIZATION IN THE SOUTH

The organization of the South by the South reached its zenith during the Civil War. Subsequently it has seemed sapped of the will to organize, has resented organization from without, and has not infrequently exported its potential organizers. Yet, in one respect, organization has been forced upon it. As water conservation has been imperative in the semi-arid south-west, so flood control has been necessary in the humid south-east. Eric Linklater's Don Juan was swept away in a Carolinan flood, and the barrages which stem the 'Plainsland' rivers aim to reduce the threat of periodic submersion. Canalization for transport brings automatic control to such rivers as Alabama's Tombigbee which, with the Black Warrior tributary, serves the Birmingham area. Major rivers, however, become an inter-state responsibility. Thus, federal control was already imposed upon the lower Mississippi in the last century. It is the military responsibility of the U.S. Corps of Engineers, who strive to contain its majestic meanders within their levees and to ease its passage to the Gulf. Elaborate floodwater outlets have been constructed to Lake Pontchartrain, while an alternate scheme to hasten the process of river capture and divert the Mississippi's main waters over Port Morgan to the west is debated. The great burden of silt carried to the Gulf is deposited in North America's most impressive delta. While the river has adequate depth for the largest ocean-going steamers at New Orleans, it suffers 20-foot shallows in the delta 'passes'. Higher upstream there is increasingly strict control of Mississippian tributaries. Renaissance of river-borne commerce as well as flood control lie behind plans to clear a 9-foot channel to the Oklahoma border along the Arkansas River.

Need for water control lifted inter-state enterprise to a new level in the Tennessee Valley during the years following 1933. The Tennessee Valley Authority, a federally sponsored project, called for long-period, large-scale, regional organization. Seven states were involved in five objectives—flood control, irrigation, soil management, generation of water power and provision of navigational facilities. In the process new farmland has been created and old restored, new industry has been established and old revitalized, and a new mental outlook has been engendered.

While the T.V.A. illumined the dark years of the depression, the war years gave rise to stock-taking by the Southern Regional Council. Its balance sheet of resources still holds good. The credit side consists mostly of physical resources—the rich agricultural potential of the long growing period, the mineral resources (petroleum in Louisiana as shown on Figure 40, iron in Alabama, bauxite in Arkansas, phosphates in Florida, and widespread natural gas), the wealth of the woods, the energy resources of Appalachia and the Piedmont. The debit side consists principally of human features—the low standard of health, the inferior literacy, the entrenched dichotomy in education, the shortage of trained men and women, the restrictions upon coloured labour, the low wage rates, the frequent emigration of the more enterprising elements and the depressing effects of urbanization.

There are two ultimate problems in the Old South—the one economic: the other social. The problem of the economy is dependence. The technological and financial dependence of the South on the North is persistent. Industrialists must commonly solicit aid in the North for the expansion of their plants. In the process, industry remains a white-man's business. Farmers must ultimately derive loans and mortgages from the same source, so that the direction of farm development is partly controlled by the same absentee landlords. The memory of exploitation by northern 'carpetbaggers' and 'scallywags' following the Civil War dies hard. The problem of society is the dichotomy of white and coloured people. While there is no single or simple solution to either problem, the speed and intensity of change in the South tend to bring economy and society into increasingly frequent conflict.

The Old South is not without its ideals, but they are those of a not especially progressive rural society. Sixty years ago, Henry Grady described his vision of this Utopia:

> There shall be breaking the day when every farmer in the South shall eat bread from his own fields and meat from his own pasture and, disturbed by no creditor and enslaved by no debt, shall sit among his

IX THE NIAGARA PENINSULA, ONTARIO

This is one of Canada's critical settlement areas. It lies along the frontier of the Niagara river and rises from the terraced plains of Lake Ontario, occupied here by soft fruit orchards and vineyards, to the limestone escarpment above which extends a provincial park. The Battle of Queenston Heights was fought here during the War of 1812. The transect (Fig. 52) follows approximately the same section. Hydro-electric power installations can be seen adjacent to the Niagara river gorge and new highways have intruded upon the fruit orchards on the dip slope

X RIVERSIDE SETTLEMENT BELOW QUEBEC CITY

This is one of the oldest settled areas in Canada and it lies immediately to the north-east of Quebec City. The historic pattern of land division is very persistent. The long bridge links the isle of Orleans to the mainland: the Montmorency Falls are seen to the east

teeming gardens and orchards and vineyards and dairies and barnyards, pitching his crops in his own wisdom and growing them in independence.

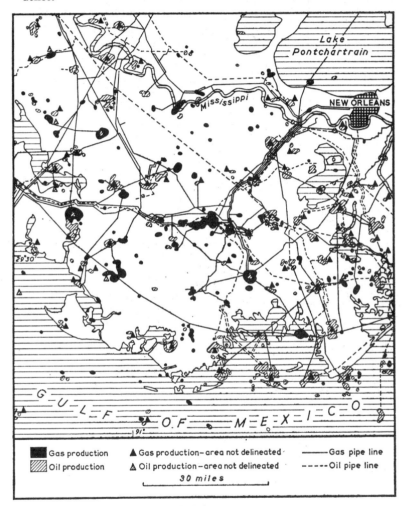

Figure 40 *Oilfields in southern Louisiana*

The elements of the landscape attributable to oil are summarized on this map for a part of Louisiana. Offshore as well as mainland resources and installations are identified, together with the network of oil and natural gas pipe-lines (Source: *Oil and Gas Map of Louisiana*, Dept. of Conservation, Baton Rouge, 1959)

The picture is physically realizable, but it demands a fuller spirit of co-operation (as well as the maintenance of agricultural supports). The inspiration of the T.V.A. illustrates the contribution of co-operation at the highest level. The 'Men of the American Midi' react more slowly than, and differently from, the men of most other parts of the union; but they are on the move. Yet Kentucky, Tennessee, Mississippi, Alabama and Arkansas still have the unfortunate distinction of leading the U.S.A. in the proportion of rural poverty.

SELECTED READING

The most concise and entertaining study of the area is

J. F. Hart, *The southeastern United States*, (London, 1967)

This should be read beside a magnificent background study

L. C. Gray, *History of Agriculture in the Southern U.S.A. to 1860* (New York, 1941)

The nature of the 'hill-land South' is well presented in

South-eastern Excursion Guide Book, XVII Congress of the I.G.U. (Washington, 1952)

and the essential features of a sample state are summarized in

J. Gottman, *Virginia at Mid-Century* (New York, 1955)

Rural matters are regionally treated by

R. N. Ford, *The Everglades Agricultural Area* (Chicago, 1956)
D. E. Christenson, *Rural Occupance in Transition* (Chicago, 1956)

and urban matters touched upon in

J. F. Hart, 'Functions and Occupational Structures of Cities of the American South, *A.A.A.G.* 45 (1955), 269–286

Two contrasted systematic studies are

M. Prunty, 'The Renaissance of the Southern Plantation', *G.R.* 45 (1955), 459–91
R. W. Harrison, 'Lousiana's state-sponsored Drainage Programme', *Southern Economic Journal* (1948), 387–403

Industry is usefully covered in

G. E. McLoughlin and S. Robuck, *Why Industry Moves South* (Kingsport 1949)—a locational analysis based on interview techniques

J. U. van Sickle, 'Industrialization and the South', *Southern Economic Journal* (1949), 412–24

Finally, we recommend

W. J. Cash, *The Mind of the South* (New York, 1942)—a rounded study of southern attitudes

G. R. Clapp, *The TVA: an Approach to the Development of a Region* (Chicago, 1955)—reviews the machinery and mood surrounding an epic project

J. F. Hart, 'Changing Distribution of the American Negro', *A.A.A.G.* 50, 3 (1960), 242–66—analyses primary ethnographic changes

7

The Great Plains and the Problem of the American Dry Belt

A Most American Part of America—'*The Pastoral Garden of the World*'—*An Interpretation of the Great Plains*—*Water Supply and Water Control*—*The Three Faces of Farming*—*The Rise of Industry*—*The Refuge of Ozark-Ouachita*—*A Formula for Understanding*

A MOST AMERICAN PART OF AMERICA

The Great Plains are in some respects the most American part of America. The humid, wooded east has West European characteristics: the Pacific Coast has familiar Mediterranean and temperate oceanic environments. The semi-arid west—low plains rising to high plains and, beyond the Great Divide, to intermontane plains—is wholly different from them. It is open country in contrast to the closed country of the east: grass and scrub take over from timber; surface water is regionally and seasonally deficient. The Great Plains stretch from the Mexican border on the Rio Grande to the wooded fringes of the Peace River. J. W. Powell in his *Report on the Arid Region of the U.S.* (1878) first defined the 20-inch isohyet and the 100th meridian as their eastern margins. Vegetationally, the boundary lies where the big woods take precedence over grass, between the Red and Mississippi valleys in the north, and along the Missouri Valley and western slopes of the Ozark-Ouachita uplands in the centre. It is a ragged margin, with tattered ribbons of woodland strung along the entrenched valleys. The eastern edges are often picturesquely called 'the break of the plains' and the Rocky Mountain foothills define a fairly clear boundary in most areas. The Great Plains are divided administratively between Montana, North Dakota, South Dakota and Wyoming in the north; Nebraska, Colorado and Kansas in the centre; Oklahoma, Texas and New Mexico in the south.

The plains offer a continuing challenge because of their extent. They average 300–400 miles in width and about half of their area lies more than 2,000 feet above sea level. The western half, consisting of a vast debris apron at the approaches to the Rockies, is named the High Plains. They embrace outlying highlands such as the Black Hills of South Dakota, true badlands where rock crumbles to dust at the touch as in Wyoming and the 'redlands' of the south with their Red Hills and Red rivers. In Texas and New Mexico the High Plains bear the Spanish name *llano estacado* or staked plains. Texas is the largest of the states sharing the Great Plains and in it the plains meet the sea. Its Gulf Coast Plain is timbered with live oak, pin oak, elm, cottonwood and southern pine, in a belt 100–150 miles wide. Atlas maps, such as those of T. G. Bradford, which were contemporary with the days of the Texan Republic, inscribed the phrase 'Great American Desert' over the Great Plains, and Captain Marcy's expedition reported 'it would seem as though the creator had designated [the Great Plains] as an immense barrier beyond which agriculturists should not pass'. In the process of their settlement woodland man was transformed into grassland man. The pendulum of their appreciation might swing in the other direction when William Gilpin chose to call them one of the world's pastoral gardens. Settlers discovered them to be neither desert nor garden, but unpredictable country where averages meant next to nothing. To break the soil was to gamble. Yet plenty of fortunes were to be made for the plenty of hopes to be shattered. If the risk of drought could be eliminated, much of the plains could be good land. The Great Plains are one of the most American parts of America in the scale of the problems which they impose and the rewards which they promise.

The plains are a problem area in a more narrowly academic sense. Two generations of American naturalists have been bred upon them and perplexed by the origins of this open grassland. Their speculations provide some of the most absorbing literature in the broader field of natural history. It is a matter for continuing debate whether the grasslands are the original form of vegetation, whether they are animal-induced or man-induced. The analyses of the nature and needs of the plains, springing from the universities of Nebraska and Kansas, are among the most stimulating of American ecological investigations.

The plains are one of the most American parts of America in a more homely respect, for upon them was born one of the country's most widely known and universally emulated folk heroes. The cowboy is one of the ultimate human expressions of the plains, for his existence depends on grass. The cattle, which are the cause of his labour, feed upon it: so did, and to a lesser extent does, the horse which enabled his forerunners to conquer the plains. The community into which the cowboy has been born is scantily peopled and has but a tenuous hold on much of the land. Most inhabitants either love or loathe the plains for this reason and there are few with lukewarm feelings. The revolvers which they carry are no toys, but weapons of defence against a wild nature. The high-heeled leather boot is an essential protection against snake and scorpion: the 10-gallon hat against the bright sun. The life of the cowboy has changed as the West changes, but he is indispensable while the West remains cattle country. The Great Plains alert all five senses—and the sixth as well. In one of his few geographical sorties (to the Royal Geographical Society in 1914), Rudyard Kipling summed up these semi-arid lands in

> . . . a heart-searching little motif of five-notes—horse, old saddlery, coffee, fried bacon and tobacco . . . that can carry a man down from high dry camps in the Selkirks . . . over red spicy dust and dead white dust through the scent of sage brush and sharp peppery euphorbias, down to the torrid goat-scented south, where fried beans, incense and the abominable brassy smell of pulque will pass him on to the forlorn brood of mangrove foreshore.

Half a century later there are additional motifs. Among them is the smell of crude oil as it gushes to the surface, the multitude of scents from the cracking plants and refineries, the exhaust of the automobiles which tie the land together and of the tractors that turn its sod, the smell of fuel which heats the basemented homes in the north and which cools the basementless dwellings of the southern plains.

'THE PASTORAL GARDEN OF THE WORLD'

It is estimated that 38 per cent of the U.S.A. would support natural grassland and 14 per cent desert shrubs. Four principal grassland complexes were originally distinguishable, although they displayed many variations within each complex and have been much altered through grazing and cultivation. The low eastern plains of the Dakotas, Nebraska, Kansas and north-west Missouri supported the tall grasses common to the grovelands of the Middle West—the aristocratic 'bluestems' (*Andropogon* and *Agropyron*). The tall-grass prairie covered 16 per cent of the land area. Beyond the transitional mixed-grass prairies, there were and are the short-grass plains which form the heart of the grassland region. Principal constituents are the blue gramma grass (*Boutelous*), the Buffalo grass (*Buchloe*) and the wheat grass (*Agropyron*). The short-grass plains cover some 14 per cent of the land area. Next, there are the resistant 'desert' grasses of the south-west, the nutritious bunch grasses which provide both winter hay and summer grazing, and the mosquito grasses (*Hilaria* species). Finally, there are the semi-desert shrubs, e.g. the big sage-brush (*Artemesia*) and creosote bush (*Larrea*), covering 14 per cent of the land area. The semi-desert lands are characterized by a limited vegetative covering and by a restricted plant variety, but they include many so-called 'forbs' and browse plants, which are quite palatable to grazing animals.

Precipitation increases from west to east, and effective precipitation diminishes rapidly to the south as evaporation increases. Vegetation belts therefore cut diagonally across precipitation zones (Figure 41). The so-called short grasses which will thrive with 14 inches of rain in Montana, require 17 inches in Colorado and 21 inches in Texas. Grasses display a range of morphological adaptations to these arid conditions. Some evade drought through growth restriction by adopting a bunched form. Some resist through deep and elaborate root development, water-storage methods and xerophylous features such as thickened cuticles; some escape through a burst of energy and quick seeding after rain; others endure through pro-longed seeding mechanisms. Because the grasslands stretch through extended latitudes varied daylight conditions also evoke a sensitive response.

A mistaken image of uniformity and monotony has been built around the Great Plains. In detail there is great diversity. In addition to larger herbivores such as the buffalo, a variety of lesser browsing animals is native to the range lands. These burrowing creatures and rodents, which give to North Dakota its nickname 'The Flickertail State', are matched by a rich insect life. Kansas alone claims nearly 200 species of grasshopper. Fauna, flora and insect life are delicately adjusted in their mutual rela-tionships and form an assembly known to American ecologists as the 'grassland biome'. Expansion and contraction of flora in any locality is inseparable from changes in the distribution of the fauna.

Uniformity as a feature of the grasslands is also challenged by climatic variability. The pastoral garden of the world is far from a constant concept. There are marked seasonal variations in the appearance as well as the stock-carrying capacity of the land. The so-called 'mesquite-desert grass savannah' of Texas, shrivelled brown for much of the year, breaks out in the spring. In April the roadside verges are a herbaceous border brilliant with the bluebonnet (state flower of Texas), the vermilion

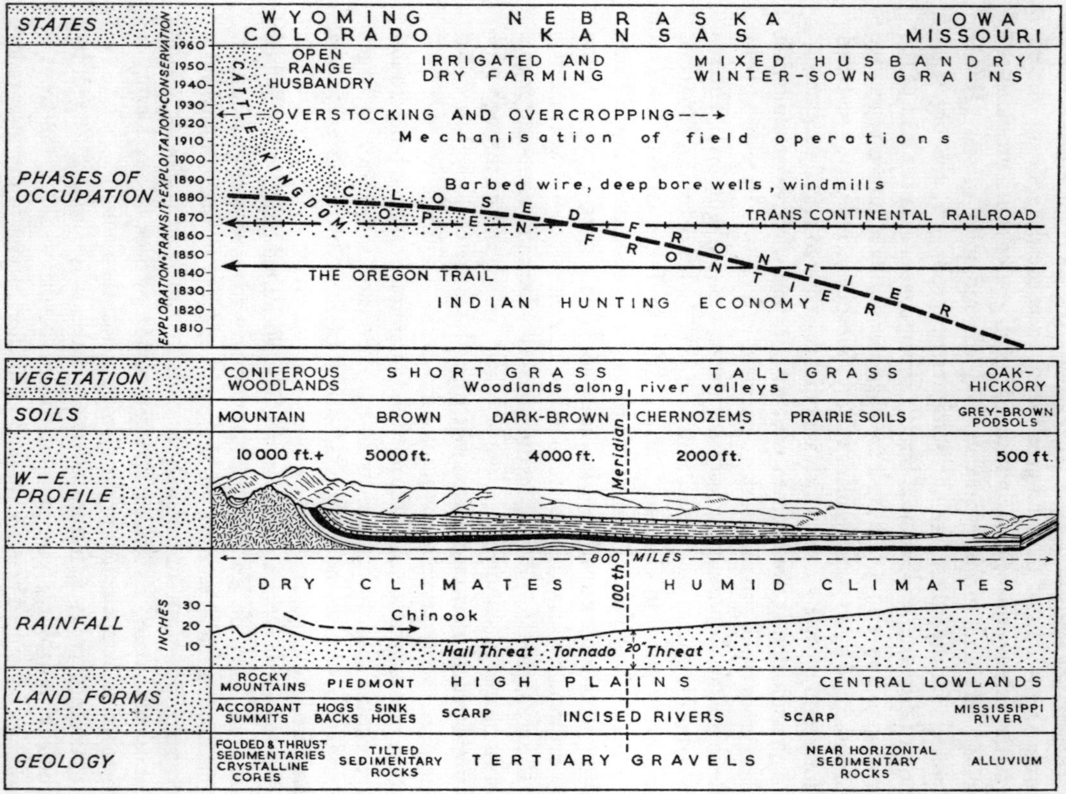

Figure 41 *A Transect through the Great Plains*

Indian paint brush, yellow coreopsis and pink primrose in grasses green as those of the Emerald Isle and oak groves as lush as those of Sherwood Forest. No wonder J. B. Priestley called his Texan sojourn *Journey down a Rainbow*.

Cyclical variations in precipitation also disturb constancy in the distributions of flora and fauna. Expansion and contraction of dominant species follow the recurrent cycles of drought. Perhaps men disturbed the grassland biome with their overgrazing and their overstocking to give rise to the storied years of the dustbowl and John Steinbeck's legend of *The Grapes of Wrath* (1939). Yet there is abundant archaeological evidence for the dry cycles. Eleven years after the great 1935 blow-out of soils Kansas was producing higher crop yields than ever and Oklahoma was a musical-comedy land. Perhaps the years of occupation have not been sufficiently long for men to adjust themselves to the behaviour and rhythms of the Great Plains.

AN INTERPRETATION OF THE GREAT PLAINS

The reactions of settlers to this unfamiliar land is the theme of one of the most remarkable studies written about North America—Walter Prescott Webb's *The Great Plains* (New York, 1931). The book, sub-titled 'a study of environment and institutions', is the work of a historian who lives in the plains and who has detected in them one of the continent's most distinctive natural regions. In essence, Webb saw the land as 'nakedly simple in its physiographic features' and upon it he observed (or lived among people who had observed) the rapid evolution of human society. For the most part, people came to it from 'a humid climate, with instruments adapted to a humid climate'. In order to solve the physical problems which faced them in the plains, 'they were compelled to modify these instruments, whether tools, weapons or legal institutions'. It is a deterministic investigation, underlining the manner in which geography has influenced the course of regional history. At the same time, it is a uniquely rounded study in which the author insists upon the unity of life in the plains and the impossibility of understanding any of the threads of its history in isolation. Because of the unitary approach, few more convincing regional studies have been written.

Fundamental to the occupation of the plains were the adaptation of old techniques and the adoption of new. The plains were inhabited by tribes of predatory Indians, whose hunting skills had been transformed by their acquisition of the horse. The white man mounted with a rifle was a poor match for the Indian mounted with a bow and arrow. Plainsland warfare was revolutionized by Samuel Colt's invention in 1835 of the pistol with a revolving chamber. Adoption of the six-shooter in place of the single-shooter facilitated military occupation of the plains.

Agricultural occupation awaited the full impact of the revolutions in industry and transport. Prairie schooner, pony express or imported Smyrna camel were inadequate for carrying the heavy and abundant technical equipment needed from the east and for exporting the products of the cattle kingdom which was emerging in the West. The railroads could carry timber to replace the sod house which was the historic plainsland counterpart of the log cabin. They could bring in steel boring apparatus to give access to underground water. They could transport the components of cheap windmills such as the Jumbo and Battle Axe to use the proverbially high winds and to replace the windlass and well-poles of the humid east. They could carry barbed wire to bound properties, to impose order upon open ranging and to precipitate the brief fence-cutting war when traditional trails were intercepted. Cheap transport also brought in essential farm machinery, with interchangeable parts.

The agricultural occupation of the Great Plains was different socially as well as technically. The plains were peopled by free land settlers, and the Homestead Act of 1862 provided new conditions of occupation. As settlers pushed into the semi-arid west it became necessary to work with larger units of land. The 640-acre section was subject to division in the humid east: it became the object of multiplication in the West. Pasturage districts of 2,560 acres were granted. Such large tracts implied that settlers were thinly spread over the land, with all the accompanying relaxation of legal and social convention. The dry lands also called for a re-interpretation of the law appertaining to water.

Adding all these changes together, Webb came to the conclusion that in passing from the woodlands to the grasslands, North American man was forced to make a profound break from his established way of life. The consequences of this amounted to the creation of an institutional 'fault line' at the meeting ground of the two environments.

WATER SUPPLY AND WATER CONTROL

Availability of water is the major problem of the plains and of the arid intermontane beyond. It is a problem of availability on the land, beneath the land, seasonally and cyclically. The uneven distribution of water resources has many causes: for the occupation of the plains it has had many consequences. Most of the water of the semi-arid plainslands derives from related humid areas, and perhaps less than a quarter originates within them. Administratively, this implies dependence of the 'rainshadow' and 'desert' states on the 'mountain' states.

Over most of the Great Plains rainfall is deficient in volume and intermittent in occurrence. Despite relatively slight slopes, there is a high rate of run-off in flash flood and torrent because of the concentrated downpours. Water absorption and retention is relatively low because of the nature of the plainsland soils. The proportion of organic matter in the soils tends to diminish westwards and soils low in humus content have a low absorptive capacity. The dry and persistent winds of the plains contribute to the high rate of evaporation as well as initiating the dust storms which darken the skies. In the foothills and mountain fringes, high solar radiation in relation to slope and aspect, and a *föhn* or *chinook* effect, induce direct evaporation and reduce snow run-off. The rivers which are fed from this hill country and which commonly flow in valleys that tell of a formerly larger volume of water, not infrequently shrink in size on their eastward journey.

Water in the West is needed primarily for promoting agriculture. The U.S.A. has more than 21,000,000 acres of irrigated land and most of this lies west of the 100th meridian. Two different types of irrigation project are found. First, there is the multitude of small-scale schemes—individual or partnership enterprises; co-operative, company or local government undertakings. Most of these look to stream diversion, to 'tanks' and to the groundwater of local wells and artesian basins. They provide the simplest, cheapest but not necessarily the safest insurance against drought. They are the projects most likely to suffer from the short-period arid cycles, to be affected by pollution and to sense lowering watertables. Groundwater is a highly valued source of irrigation because of its constant temperature and freedom from silt. But if it is saline or contains chemicals in solution, expensive extraction equipment may be needed.

Large-scale projects, the second type, are usually complementary to, rather than competitive with, small-scale schemes. Most rivers traversing the plains have water-conservation schemes, and maps of the area indicate the strings of artificial lakes along their courses. Some rivers have been subjected to major engineering projects, under the aegis of state or federal enterprise and with the collaboration of all the regionally interested government departments. The upper Missouri River provides an example. Below the 100-mile-long Fort Peck Reservoir in eastern Montana are encountered the Garrison Reservoir of North Dakota, and the Oahe and Fort Randall reservoirs of South Dakota. The Boysen Reservoir on the Big Horn River brings life to the high arid basin of central Wyoming; the Elephant Butte Reservoir is central to the dry uplands of southern New Mexico, while the Falcon Reservoir is on the international reaches of the lower Rio Grande. But in the south especially the exposure of such vast surfaces of water almost defeats itself, since there is so much evaporation.

Once the settlers of the humid east encountered water shortage, new approaches to water rights came into being. The law of riparian rights, springing out of the Common Law of England, was replaced by the principle of prior application. The settler who first employed a water resource beneficially established a right to it. The establishment of state boundaries introduced further legal obstacles to the free employment of water, until such time as it was decreed that there was equality of rights at the inter-state level.

In more recent years shortage of water has directed attention to water wastage as well as to water use. At the higher level, flood control plays a major part. At the lower level, conservation techniques are vigorously debated by many schools of thought—the maintenance of cover crops, summer fallowing (especially beyond 100°W), mulching, rotations, shelter-belt planting, rationalized grazing, shallow ploughing (or even scuffling the stubble as in *Ploughman's Folly*), the destruction of water-loving shrubs such as the tamarisk. The more scientific use of irrigation water is another expression of conservation. Whereas in the humid east the criterion of production is yield per acre, in the irrigated west it becomes yield per acre-inch of water. Irrigation in itself must be carefully managed, for over-watering can leach out chemical constituents and affect soil quality.

Ultimately, the agricultural potentialities of the sunny western plains are an expression of water management. There is insufficient water to bring them all into the intensively farmed realm. Yet, given capital and initiative, there remain many areas into which the frontiers of more intensive cultivation could march. There are also many other areas where improved management could have an indirect effect upon water supply. It is only 70 years since the first of a series of National Irrigation Congresses was held in the semi-arid west to win the support of federal authority and to diffuse experiences. Achievements have been great in the interval. On the other hand, the first experiments to induce precipitation, held 70 years ago in Midland, Texas, achieved no more success than the dances of Indians to their rain gods. Scientific rainmakers are still seeking a formula with which to seed the clouds.

THE THREE FACES OF FARMING

A generation ago agricultural maps of the Great Plains neatly divided them among a series of staple crops. In the north, the spring wheat 'belt', contiguous with that of the Prairie Provinces of Canada, extended through eastern Montana and Wyoming, North and South Dakota. To the south, the winter wheat 'belt' focused upon Kansas and Oklahoma. The cotton 'belt' had expanded beyond the historic South into Texas and Oklahoma. Westwards lay the grazing lands. This simple picture has been much changed. It represented an intermediary phase not only in the adjustment of farmers to the Great Plains, but also in the geographical interpretation of the adjustment. Today the lines of definition between the regional concentrations of activity have been blurred. As elsewhere, combination of a number of crops and the closer integration of stock have taken precedence over regional specialization. The sensitive reaction to climatic hazard and the sensible drawing together of all forms of rural enterprise are at one.

The critical phase in this process occurred a generation ago when physical and economic disaster struck the area simultaneously. Deserts seemed on the march over much of the area, at the same time as an international slump in farm prices hit the unprotected farmer. Much farm abandonment resulted: much land fell into arrears in payment of local taxation. Local administrative authorities found themselves in possession of varying extents of land held derelict in their commission. Some of this land ought not to have been put under the plough, some of it had been wrongly farmed, some of it was of potential value given large-scale capital investment, e.g. in irrigation. Today both land and men are the objects of protection. Through a variety of conservation schemes, implemented at the farm level by educational as well as financial controls, the land is more scientifically treated. Uncertainties are smoothed out increasingly by means of the Agricultural Adjustment Acts, federal crop insurance (since 1938), acreage restrictions (the idle land constituting a so-called soil bank) and, in extreme need, extension of Disaster Area legislation.

Contemporary plainsland farming is classified as grazing husbandry, irrigated farming and dry farming. To a certain degree the three are integrated, but on the face of the land they are distinctive. Their character varies greatly from north to south. For example, the longer growing season and hot summers of the virtually sub-tropical south introduce contrasts through both the greater range of cropping opportunity where water is available, and the diminished stock possibilities where it is not.

Grazing husbandry predominates on the High Plains. Its extensive character is reflected in the definition of a large farm in the preamble to the Agricultural Census. In the east, a farm is large if it exceeds 1,000 acres: in the west, 5,000 acres. In the east, if it has 200 cattle or 500 sheep: in the west, 500 cattle, or 4,000 sheep. Distances between farm units are multiplied correspondingly. Where stock-carrying capacity may fall to one beast per 50 acres, light aircraft and helicopters may take over from jeeps as a second complement to the indispensable horse. In the southern Great Plains especially, the open range is the main source of forage. Management implies much more than prevention of overgrazing. It means the substitution of higher value for lower-value forage plants and the struggle against infestation by poorer grasses or shrubs. The High Plains are transitional to the foothills and passes open from them to the intermontane basins. Cattle and sheep are supplemented by goats in the south. Transhumance, by wheel as well as on the hoof, reaches a scale in Wyoming, Colorado or Montana rarely encountered in Europe even in the classical days of seasonal migration. Until the Taylor Grazing Act of 1934, federally owned land in the western states was largely treated as common land and heavily overgrazed. Livestock operators now lease this land from district advisory boards on temporary (one-year) licence or long-term (ten-year) permits. Grazing has also been formally controlled, upon the basis of established rentals and stabilization of animal numbers, through the National Forest Administration. Ranging on Indian Reserve Lands has been dealt with as a part of the Indian Reorganization Act of 1934.

Stockmen, once competing against the arable farmer, are increasingly complementary. Wintering fodder or 'finishing' rations are widely grown on irrigated land. To more familiar rotation grasses and 'sweet' clovers are added a growing range of legumes and catch-crop millets. Alfalfa provides the chief legume hay of the plains. A traverse of the North Platte Valley shows alfalfa stacks dotting the irrigated fields, the alfalfa dehydrating mills of Gothenburg and the town of Cozad which claims to be the world's largest alfalfa shipping centre. Given irrigation, such grasses as the runner-propagated Bermuda grass (*Cynodon spp.*) give the illusion of permanent grass. Sorghums, used as a green as well as ensiled fodder,

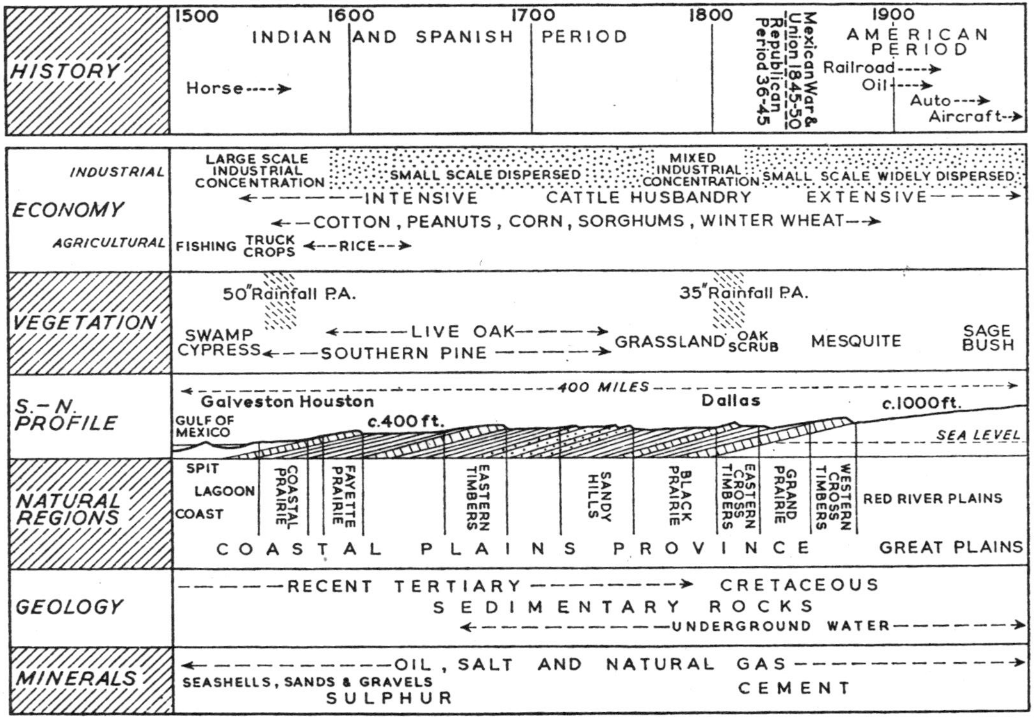

Figure 42 A Transect through southern Texas

are the second fodder crop of the northern Great Plains. Integration between corn or cotton-seed processing plants and animal fattening is close. The cattle kingdom of the west has grown increasingly self-contained and diminishingly tributary to outside areas. And in Texas, as in the plainsland areas of the Old South, it takes on a new complexion through experiment with tropical stock. There are farms, e.g. at Hungerford, which breed in spring and autumn with a view to selling 500 Brahman or Brahman crosses a year. As with the rest of the U.S.A., Texas is also a heavy consumer of milk, so that dairying is of growing consequence.

Some indication of the range of crops cultivated with or without irrigation on the Gulf Coast plain is given in Figure 42. The transect indicates rice fields wet or dry-planted (fertilized and even sown from the air), peanuts, wheat (which ripens in May), coarse grains and cotton which have replaced the cleared timber, the drained swamp, the ploughed-over scrub and thorny mesquite. Much of this agriculture is of a large-scale, sometimes neo-plantation character, as indicated by the distribution of settlement on Figure 43. Much is tied in closely to the managerial section of the industrial community. But there are plenty of small farms, strung along Rural Routes, the postal addresses of which exceed R.R. 1000. Farmers' Co-operative Gin Companies (cotton gin that is to say) with their rough wooden sheds and modest timber-covered livestock markets (outside almost shadeless country towns) reflect enterprise of a contrasting order. On the irrigated flood-plains of the Spanish-named rivers from the Trinity westwards, vegetable crops are intensively cultivated; while in the Rio Grande Valley itself citrus fruits rival those of California and Florida (where, as with the Texan 'pink' grape-fruit, they do not take precedence over them).

While the compulsions operating in the extremities and margins of the Great Plains have forced or favoured their particular adjustments, there remains a central area which perplexes practising farmer and experimental officer alike. The western halves of Nebraska, Kansas and Oklahoma are neither so agriculturally restricted as Texan equivalents nor so agriculturally safe as their Dakotan counterparts. They are bisected by the critical 100th meridian, and they lie in the area which has been practically and theoretically identified as an area of risk. The need for a scientific management of range-lands has made its impact and overgrazing is on the way out. The role of irrigated farming is self-evident. The least-healthy face of plainsland agriculture is that which is turned to dry farming.

Dry farming remains to a large extent specialist farming in an area of

XI THE RIA COAST OF NOVA SCOTIA

The ria entrance to the harbour of Pictou, with the Northumberland Strait to the north-east. The mixed farming country is interspersed with woodlands of mixed quality. Along the coast, sand bars protect fishermen's coves

XII THE LAURENTIAN SHIELD OF THE NEAR NORTH, ONTARIO

This tract of Shield country, showing the settlements of Swastika and a part of Kirkland Lake, is typical of the landscape at the southern end of the transect depicted in Fig. 56

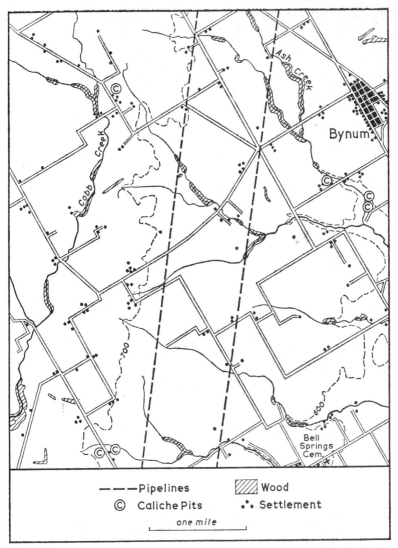

Figure 43 *Sample Settlement Features from the extensive Farming Country near Dallas (Texas)*

(Source: *U.S.G.S., 1 : 24,000 Abbot Quadrangle*)

difficulty. It is at the same time conducted against the background of farms of moderate status. It includes for example specialist grain areas in western Nebraska which still bear visual testimony to O. E. Baker's wheat belt. Wheat is still to be found growing as a continuous crop and local overcropping undoubtedly occurs. The percentage of failures is tied closely to short-period variations in aridity, to extensions of improved cultivation, or to both. Detailed investigations by Leslie Hewes in the critical marginal zone of western Nebraska have shown a decline from 40 per cent failure (1931–7) to less than 15 per cent failure (1948–54) in many counties. Failure springs from several causes. Drought is central, but winter-killing (through frost-heaving and icing as well as sharp temperature falls) intrudes as a powerful cause from the east to match hail from the west. Soil blowing is even more destructive than plant disease or insect (principally grasshopper) pest. Experiment and practice over the last generation have confirmed that summer fallow substantially increases returns under average conditions, and reduces losses under extreme conditions. Insurance on the ground as well as through the agent is expressed in practices such as stubble mulching, strip-cropping (across the direction of the wind), deep-furrow drilling and more widespread use of winter-hardy grains. All such practices suggest that facing the dry lands without irrigation and without animals is today more in the nature of an exercise in insurance than a gamble with natural hazards.

Finally, there is a growing integration between farming and other branches of the economy. Subsistence farmstands are found, but hobby farms multiply steadily. As in other parts of the world, the industrial magnate finds in a model ranch an additional prestige.

THE RISE OF INDUSTRY

Industry in the Great Plains is very unevenly distributed. It is associated principally with the refinement of raw materials and it is concentrated in the 'Southland'. Indeed, there is a primary economic as well as physical contrast between the southern and the northern plains. It is related closely to climate with its effects on agricultural opportunity and to the presence or absence of oil and natural gas. Uneven distribution of industry has resulted in an uneven distribution of population. As a consequence, the Great Plains have some of the most thinly peopled and, in their main metropolitan areas, some of the most densely peopled parts of the U.S.A.

Three principal types of industry occur. The most widespread are those associated with the products of the soil, but they are often of modest scale. They only attain significant stature and diversity where the consuming market is big and where there is a substantial range of locally produced raw materials. Industries based upon mineral refinement are dominant in Texas and Oklahoma; though extractive industries, such as the Butte and Helena copper mines, are still associated with Montana (if not physically with the plains). Tributary industries, which are often branch plants, reflect the size and wealth of the local market. Texan industry has its own ancillary plants beyond the state boundaries.

The industrial wealth of the south-western plains is built on oil. No state has more oil wells than Texas: it is unlikely that any has greater reserves. In few states is oil so widely distributed, and producing centres occur in most counties. The principal oilfields are the Gulf Coast field, the Mid-Texan field and the Mid-Continental field. Commercial development of the oilfields of the south-west (including those of Oklahoma and Louisiana) belongs to the last 60 years. Texan oil was first struck in 1894 at Corsicana. Oil occurs at varying depths from several hundreds to several thousands of feet. It extends offshore into the Gulf of Mexico (Figure 40). *The Oil and Natural Gas Year Book* lists by county the changing number of active wells. In general, oilfields do not make a great impact on the landscape. The pin-cushion distribution of pumping machines (which may work under their own pressure or may require an external source of energy), and of oil derricks (which may be of aluminium as well as of wood), is found offshore on the sea bottom, in the wooded Louisianan *bayou*, the coastal plain rice or cotton field, and the panhandle's open range of brushland (Figure 44).

As a natural resource, oil has the advantage of easy mobility. The three great south-western oil-producing states are threaded with both covered and uncovered 12–20-inch pipe-lines that convey the oil from the point of supply to the points of consumption, refinement and export. Those which extend to the industrial north-east take the Mississippi in their stride. The principal centres of refinement are not usually at the points of production; and a number of producing centres are usually tributary to a refinery. Among inland processing centres, Tulsa (Okla.) claims to be the 'oil capital of the world', while the state capitol in Oklahoma City has symbolic derricks standing on its lawns. But it is on the coast especially around Galveston Bay that refining reaches its fullest expression. Crude, semi-refined and refined products move out of Baytown harbour, along the Houston Ship Canal by tanker to overseas and east-coast domestic markets, and by barge via the intra-coastal waterway and

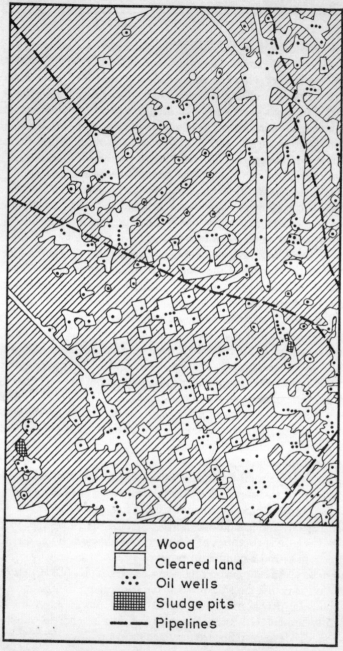

▨	Wood
▢	Cleared land
⋮	Oil wells
▦	Sludge pits
– –	Pipelines

Figure 44 *An Oilfield from Conroe* (*Texas*)

In coastal Texas, oilfields are developed against the background
of heavily wooded country. This map illustrates the pattern of

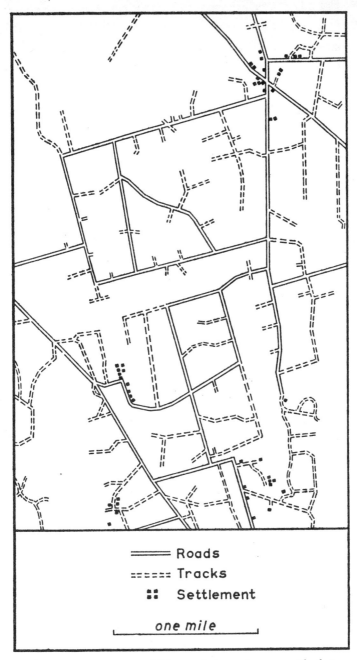

====== Roads
:::::: Tracks
:: Settlement

one mile

clearance in response to drilling and the settlements with their interlinking roads (Source: *U.S.G.S.*, 1 : 24,000, Conroe Quadrangle)

the Mississippi to the interior market. Other major centres of petroleum export are Corpus Christi and Brownsville. The refineries are basic to a whole group of manufacturing industry, with petro-chemicals predominant.

Natural gas is also associated with oil-bearing rocks. Together with petroleum it now makes a more substantial contribution to the American pool of energy than coal and water power. Both flow more cheaply than electricity can be transmitted. Texas, Oklahoma and Louisiana have the greatest regional concentration of resources. Natural gas is converted into energy, is exported by pipe-line, is liquefied for overseas shipment, and refined to produce similar end-products to petroleum. Carbon black, a component of the motor tyre industry, is also a derivative. Coal exists in the Great Plains, but it is associated with the north-west rather than the south and is of limited regional importance.

Cheap petroleum and natural gas have given an added value to other resources. Petro-chemicals are a natural industrial offspring; but the abundant brines and alkali, the rich sulphur deposits, the remarkable shell banks with their lime components, and the underlying rocks are all drawn into the industrial complex. The great chemical plant at Freeport (Texas), employing more than 5,000, owes its location to a combination of available resources—natural gas and petroleum, the high magnesium component in the Gulf seawater, salt, alkalines, sulphur and the fresh water of the Brazos River. As with so many of the industrial plants of the south-west, Freeport has a bulk production—agricultural ammonia, soil fumigants, chlorine, glycerine, magnesium and synthetic fibre elements.

Texas, with adjacent Louisiana, is the world's largest producer of sulphur. At such sites as Boling Dome, near New Gulf, Moss Bluff Dome near Houston and Spindletop Dome near Beaumont, the mineral is siphoned out in solution from depths of 300 to 2,000 feet. Subsidence around the site gives some indication of the volume extracted and yellow sulphur vats, 50 feet high and containing 500,000 tons, are broken up for shipment by hopper car and barge. The chemical city of El Paso, 700 miles to the west, is a major consumer.

The most impressive array of heavy industry is found along the Houston Ship Canal (Figure 45) between the San Jacinto monument (a column loftier than that erected to George Washington) and the turning basin. Here, the various endowments of the southland can be smelt as well as seen. Grains are milled and stored; mainland and Caribbean molasses are refined; vegetable oils are extracted; commercial fertilizers and acids are prepared from local sulphur and imported phosphates; meat

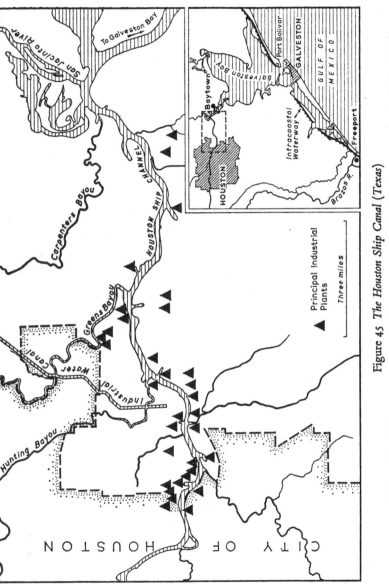

Figure 45 *The Houston Ship Canal (Texas)*

An improved channel which gives ocean-going shipping access to the city of Houston, and which provides one of America's most heavily industrialized waterfronts. The sites of the principal heavy industries are indicated (after *Houston Chamber of Commerce Map*)

and bones are processed; Portland cement and poultry grit are manufactured from Bay oyster shell; softwood timbers are converted into
paper, pulp, wallboard and alcohol by-products; bulk chemical products
range from insecticides and weedkillers to barge-shipped glycerine.
Petroleum refining, with bulk production of synthetic rubber and plastics,
remains the giant in an industrial complex where the emphasis is upon
size—from the world's largest fertilizer plant to the world's largest
tetraethyl lead producer. Ordnance and munition plants are related to the
chemical and metallurgical factories. Houston's 1,600 industrial establishments claim nineteen out of the twenty basic American industries.

In a countryside with distances more extensive and more open than
those of the Middle West, and with a population of equal wealth and
mobility, it is natural that transport industries should be attracted to the
south-west. Aircraft manufacture, e.g. Fort Worth and Dallas, benefits
from extensive areas of flat land (for testing grounds and workshops) and
good flying weather. Partly in order to serve such factories and partly to
supply the immense constructional demands, high grade electric steel
and ferro-alloy furnaces are found in Houston; though for transport
industries many parts must be imported from the north-east.

To a greater or lesser degree, the Great Plains are food-producing, and
larger or smaller food-processing factories are widely distributed. Food-
processing industries are dominant in the northern plainlands, especially
milling and meat-packing (e.g. at Omaha or Kansas City). The occupational census shows that Texas also has more people employed in food
processing than in any other single branch of manufacture. It is not only the
scale of expected activity, e.g. corn refinery in Corpus Christi, but also that
of more sophisticated foodstuff industries which distinguishes the south.
Dallas, for example, has 2,000 persons employed in a single candy factory.

The Great Plains have access to the sea through Texas. The range and
bulk of raw materials produced in the hinterland of its Gulf coast have
favoured the rise of major exporting harbours with all the servicing
activities associated with shipping and commerce. The importance
ascribed to the coast is measured by the degree of control imposed upon it.
The well-defined and maintained intra-coastal waterway gives a protected
deep-water route from Brownsville to New Orleans. The Houston Ship
Channel provided a 57-mile link from the biggest city of Texas to the
waterway at Bolivar Roads in 1915. Specialized ports and harbours have
grown in response, from the diverse tanker ports to the cotton wharves
of twentieth-century Galveston which has fallen heir to the export
function of nineteenth-century Savannah.

Industrial transformation of the south-west takes place in a setting which contrasts with that of the historic north-east. First, it is set upon a plainland, where physical obstacles are unlikely and where place-name elements incorporating hills and ridges are figments of the imagination. Secondly, it is set in a hot country which commonly avoids the impact of winter and where the creation of artificial climates in home, factory and means of transport is the accepted order of the day. Thirdly, it is developing in an area which is unencumbered by obsolescence or with legacies from the past. Such a city as Houston might be an ideal playground for those who would test the models of Lösch in a real-life setting that most nearly approaches his fiction. Here is a surface as near flat as possible, free from artificial restrictions upon site development and infinitely versatile in available resources.

THE REFUGE OF OZARK-OUACHITA

Between the plains, the Middle West and the South there intrudes a broken highland which stands in detachment from all three. The Ozark plateaux dominate the half of Missouri south of the Missouri River. They are complemented by the Boston Mountains in northern Arkansas and, south of the Arkansas River, by the Ouachita Mountains which extend westwards into Oklahoma. This much dissected upland area, rising to heights of more than 2,000 feet, is difficult of negotiation owing to sharp variations in relief. It has generally uninviting soils, it is heavily wooded and altitude gives to it a harder climate than those prevailing in its surrounding plains and valleys. The Ozark-Ouachita upland had already been dubbed a backwoods land by the beginning of the century.

It is sufficiently extensive to retain tracts in which the local economy has stagnated and in which old customs and practices have fossilized. It has become an area of population outflow. But much has been done to arrest its decline and to resuscitate its economy during the last generation. Overgrazing of its open range lands has been largely halted. The denuded oak groves and pine woods are being restored; although much land supports little more than dogwood thickets and redbud scrub. Reservoirs have been created for water control and power, e.g. the Lake of the Ozarks on the upper reaches of the Osage River and above Jefferson City, and Lake Table Rock behind Bagnell Dam. New freeways, modern counterparts of the historic trails, penetrate into the upland interior. They complement the winding hill-top roads which pick out the impressive

Burlington Escarpment, and which link the cleared land above the thick brush and wood of the valleys. The freeways, slashing open the ochre-coloured rocks, invite a lively tourist industry. For this is a country of limestone caverns, of rivers that play hide and seek, and of springs (both hot and cold) that have focused settlement at Springdale and Springfield, Siloam Springs and Hot Springs. Pockets of enterprise complement features from the past. Mules are tethered and goats graze beside unpainted shacks (in a country where paint is a criterion of prosperity), but a mile or two away may be found a town such as Berryville ('Turkey Capital of the World') or the intensive vineyards of Italian immigrants, around Tontitown (Ark.). Along the northern edges of the Ozarks there have been more comprehensive schemes to anchor population to the valleys. Elsewhere, as in the zinc- and lead-mining area around Eureka Springs, in leather industries and corn-cob pipes, old-established activities are sustained.

A FORMULA FOR UNDERSTANDING

The Great Plains have been neither assessed nor assimilated by the rest of the continent. Most of the plainlands were staked out in the latter half of the last century, and the entry has been largely of a precipitate character—a rush for precious metals, for oil, for strategic minerals, as well as simply for land. Three phases of land occupation might be identified—the first of enforced co-operation with environment, the second of environmental exploitation, and the contemporary phase which balances the achievements and failures of these two earlier attitudes to the land. The assessment of the Great Plains is incomplete for at least two reasons. First, their agricultural potentiality is regionally both greater (through new crops and new techniques) and less (through the growing competition for and restriction of water) than was formerly believed. Second, their industrial development is as uneven prospectively as actually. Three quarters of the plains are probably underdeveloped from an industrial point of view: both scenically and cartographically, they suggest an industrial 'desert'. The remaining quarter, parts of which give the superficial illusion of overdevelopment, is still probably not at the peak of its achievement. All development, as in the intermontane plateaux to the west, is overshadowed by water availability.

Much of the Great Plains region lacks an air of permanency in its appearance and depth in its occupation. Population is much on the move, with a flow away from the north-west and a migration into the south-west. Only where the American culture touches the Spanish is there a sense of the fourth dimension. The past emerges in the faces of people along the Mexican borderlands and in the assiduously restored ecclesiastical monuments of, for example, San Antonio. Impermanence is related to insecurity. The human hazard of the Indian can be recalled from living memory: all farmers and a good many others will have experienced the impact of climatic caprice. The battle for security is waged through capital investment in water control and crop insurance. It is also pursued through gravimetric and magnetic sounding devices which enable geologists to predict more satisfactorily than diviners with their 'witching sticks' the volume of underground resources. As a result, new settlements can be planned according to the life expectancy of mineral deposits.

There are many ways of explaining the content and nature of a piece of country. W. P. Webb offered one interpretation of the Great Plains, and although it may be replaced it cannot be disregarded. It has continuing relevance in the hierarchy of values, for appraisals show persistent differences from those in other parts of America. In the oil country, fuels reach their minimum assessment: in the dry belt, green things are so appraised that a 40-foot tree can send a house-lot soaring in price. Brief experience of the semi-arid realm has also taught the virtues of selection. In all forms of adjustment to the land, the art of the possible is more refined than in most parts of America.

SELECTED READING

Much of the literature touching upon the Great Plains is complementary to W. P. Webb (*The Great Plains,* New York, 1931), and to F. J. Turner, *The American Frontier* (New York, 1920)

A good readable historical account is given in
  John A. Hawgood, *The American West* (London, 1967)
An unusual book about a book is
  F. A. Shannon, 'Appraisal of the Great Plains', *Social Science Research Council, Critiques of Research in the Social Sciences, III* (New York, 1940)
Cartographic study in depth for a single state is provided by
  R. W. Baughman, *Kansas in Maps* (Topeka, 1961)
Grass and grazing are themes which may be pursued further in
  J. C. Malin, *The Grasslands of North America* (Lawrence, Kansas, 1948)
  U.S.D.A. Yearbook, *Grass* (Washington, 1948)

A good sociological appreciation is

C. F. Kraenzel, *The Great Plains in Transition* (Norman, Olka., 1955)

and more scientific backgrounds to grassland husbandry are provided by

A. W. Sampson, *Range· Management, Principles and Practices* (New York, 1952)

M. Clawson and B. Heed, *The Federal Lands: their Use and Management* (London, 1958)

The nature of the lands of risk is analysed in such sample papers as

L. Hewes, 'Wheat Failure in Western Nebraska, 1931-4', *A.A.A.G.* 48, 4, 375-9

L. Hewes and A. C. Schmieding, 'Risk in the Central Great Plains', *G.R.* 46 (1956), 575-87

See also

E. Joan Wilson Miller, 'The Ozark Culture Region as revealed by Traditional Materials', *A.A.A.G.* 58 (1968), 51-77

E. W. Kersten, 'Changing Economy and Landscape in a Missouri Ozark Area', *A.A.A.G.* 48 (1958), 398-418

8

The Pacific Slope and the Problem of Detachment

THE NATURE OF ISOLATION

The Far West, to which is tributary the greater part of the Basin and Range Province, is characterized by detachment from the rest of the country. From the Bitter Root Range between Idaho and Montana in the north to the southern Rockies of Colorado, successive ramparts divide the Pacific and Atlantic drainage. Between the eastern Rockies and the north-western Cascades and south-western Sierra Nevada, lies the so-called Basin and Range Province. It has many common features of landform and climate throughout its entire length. A zone of weakness—extending from the curiously unknown Gulf of California, through the Great Californian Valley, the Willamette-Cowlitz Valley to Puget Sound—is sandwiched between the western edge of the Great Basin and the Coast Ranges. The form of the coast reiterates the strong longitudinal graining.

The very magnitude of the Cordillera engenders a sense of isolation. Between the Sierra Nevada and the Rocky Mountain National Park, the Cordillera is twice as wide as in British Columbia, while the feeling of detachment is exaggerated by its broad stretches of wasteland. Interposed between the Pacific fringes and the Great Plains is so much unoccupied and barren country that it might with some justification be called America's 'Empty Quarter'. From Lake Tahoe on the Californian border to the Wyoming state boundary is 500 miles as the condor flies—as far as from New York City to Detroit. The carefully created and maintained oases of settlement within the empty land only serve to throw into relief the extent of the barren area.

Pacific America, separated from the east, is also divided within itself. In the broader sense it tends to fall into two halves. 'The Pacific North-West' has its own unity about the diversified drainage area of the Columbia-Snake river system. It embraces the states of Washington and Oregon, with inland Idaho. A complementary, though less unified, southern segment looks to the seaboard state of California, with Nevada, Utah and Arizona as a hinterland. Half of this is an area of interior drainage, commonly identified as the Great Basin: the other half drains by the Colorado River system, which with its exaggerated erosional features is an obstacle rather than aid to unity.

The main settlement areas of the coast are also divided from each other. The Willamette-Cowlitz Valley is separated from the Central Valley of California by the resistant mass of the Klamath Range, with Mount Shasta exceeding 14,000 feet. The Los Angeles coastal plain is detached from the land-locked Central Valley and from the Salton trough. The valleys of the attenuated coast ranges repeat on a smaller scale the theme of detachment. Many are *cul-de-sacs* and most are deeply entrenched: the Spanish word *canyon* is used indiscriminately for almost all of them. Finally, the Pacific does little to unite all this. While the Atlantic seaboard with its many inlets and estuaries offers a multitude of harbours and anchorages, the Pacific seaboard, with its narrow and discontinuous coastal 'plain', its broad beaches and rolling breakers, is void of protection and offers meagre invitations to shipping. Exceptions are San Francisco Bay, Monterey and Humboldt Bay (with its redwood sawmills and drying yards). Until the cutting of the Panama Canal, the sea route was by way of Cape Horn—in itself a deterrent to contact.

The range of contrasts between the Mexican and Canadian borders is different from but no less striking than that between Florida and the State of Maine. Climates embrace the Mediterranean and temperate maritime associated with the west coast of a continent. Isotherms are less vital controls than isohyets, for most is semi-arid country and vexed by water problems at large and in detail. The area contains some of the most historically remarkable and contemporarily elaborate schemes for water control in America. Earlier, they were private and small-scale: today, they are increasingly public and large-scale. Over the greater part of the Pacific slope, the irrigated lands are the principal source of food and rural income. No group of people have been more successful in turning brown lands green than the Mormons, and the very isolation of the Salt Lake shores was for them an attraction. Their Utah experiences have been of value throughout the Basin and Range Province. The Mormons have made their contribution in financing mutual non-profit-making companies as well as in practical field installations.

With the rise of new constructional techniques, large-scale publicly owned schemes of a multi-purpose character have taken precedence. In the south, the Colorado River became the centre of attention. The Hoover Dam, 700 feet high and impounding Lake Mead as the largest single reservoir in the U.S.A., serves to supply power as well as to conserve water. The utilization of the Columbia and Snake rivers represent the heaviest investment in any western river system, while the Central Valley Project of California is a venture unrivalled in its ingenuity.

Many legal issues over water rights have resulted from the continuous rise in the value of water. The principle of riparian ownership generally supplants that of appropriated ownership. In such an area as the Imperial Valley, looking to the Colorado River and at the meeting ground of Mexico, California and Arizona, water rights reach a peak of importance.

The competition between rural and urban needs becomes the more acute where every square yard of lawn must be sprinkled daily. Nowhere has the struggle been harder than in Los Angeles. For 60 years it has sought solutions. In 1903-4 it acquired rights along the Owens River, east of the Sierra Nevada, and by 1913 the Los Angeles aqueduct was completed. This life-line, employing gravity flow, was subsequently pushed north to the Mono Lake, east of Yosemite National Park: in 1941, a Colorado River Aqueduct was completed; a 500 mile pipe to the Feather River gorge is the longest link. California views with as much interest as Israel the possibility of transforming sea water into fresh water.

California's physical structure gives rise to marked climatic differences which are repeated over short distances. Seaward- and landward-facing slopes have striking vegetational contrasts. Two hundred miles from Sequoia National Park, with trees the height of Big Ben, is Joshua Tree National Monument with its groves of tree yuccas. White peaks with conifers deformed by the weight of winter snow look towards sun-baked valleys such as Imperial Valley, with its feathered date palms. In the southern Sierra the gamut from tundra to tropics may be encountered with all its consequences for vegetation and wild life.

While contrasts have exhilarated, isolation has not usually stultified development. From the human point of view the Pacific coastlands offered more agreeable settlement opportunities than anything between them and the humid east. California favoured the Mediterranean farming of the missions, nurturing vine and olive, wheat and pastoralism. West Europeans who came to sieve gold often stayed to turn the land of this suncrop country. As a result, no part of the U.S.A. has a greater range of exotic plants to vie with the native.

The mountain coast reaches a climax in Alaska, the neglected northern half of America's Pacific seaboard. Peninsula and island are contrasted in political evolution and economic development with the southern half. The problems of the Pacific North-West are exaggerated to a maximum in this corner of the continent.

THE CALIFORNIAN FORELAND

Few states encourage the use of superlatives more than California. It has the highest peak (Mount Whitney, 14,501 feet), the deepest depression (Death Valley, 276 feet below sea level) and the world's oldest living things (2,000-year-old sequoias). Its population is growing more rapidly than that of any other state: in few states has the urban area expanded more rapidly in the last decade. It is at once an architect's paradise and nightmare. Yet beside its extraordinary contrasts in snowfield and hot spring, in antique pueblo and contemporary Hollywood, stand equally curious anomalies. Its resource potential is modest and, until the development of oil and natural gas, its energy supplies were seriously deficient. It is the 'bonanza' state, where precocity of development has offset limitations. 'Extremes, extravagancies and eccentricities' characterize it.

California is divided between the coast ranges; the interior lowlands of the Great Valley and of the Salton trough oriented to the Colorado River delta; and the Sierra Nevada, extended northwards into the Cascades.

The essence of California is distilled in the coastal valleys (Figure 46). In longitudinal profile, in transverse profile and in latitudinal succession their natural vegetation and land use reflect an intimate adjustment to landform and local climate. South of San Francisco Bay, contemporary harmonies have historical overtones and undertones. Indian names are overlaid with Spanish names: both are blended with a later cosmopolitan toponymy.

In longitudinal profile the inner reaches of the valleys are the most fruitful. Given water, these protected retreats yield a rich harvest of fruits and vegetables. They are suntraps which escape the diurnal fogs that roll in from the Pacific, though they may be frost hollows into which

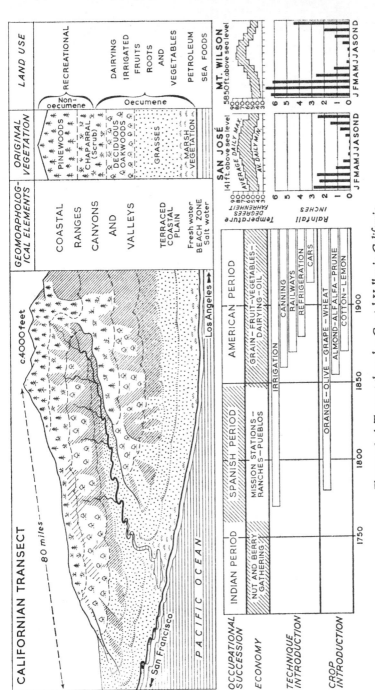

Figure 46 A Transect through a Coastal Valley in California

stagnant air drifts by night. The stench of smudge pots takes precedence over the perfume of citrus groves, midnight flames beneath the springtime blossom of soft-fruit orchards offer a sight as remarkable as glow-worms in a Kentucky cave, wind machines may whir more noisily than birds in this ornithologically limited area. The valley profile is regularly inter-rupted by windbreaks. The cypress is generally subordinate to one of twenty different kinds of eucalypts, the Australian parents of which probably arrived three generations ago. Lower down the valley, field crops replace tree crops—tomatoes, melons, lima beans, onions, sugar beet, rotation hay. Where the valley funnels into the narrow coastal plain with its marine terraces, spinach, cabbage, carrots and lucerne thrive in the damper, cooler atmosphere. Local specialities are proudly advertised—Cartersville, the artichoke capital of the world; Watsonville, the straw-berry capital; Salinas Valley, the salad bowl of America. Intensive cultiva-tion is matched by conveyor-belt harvesting. Broad beaches, barred across estuarine lagoons and beaten by a heavy cold surf, offer some crustaceans and the univalvular abalone. Offshore waters are relatively rich in fish. The 'Fisherman's Wharves' and 'Cannery Rows' of San Francisco or Monterey, with their seafood restaurants, tall tales of sharks, faded memories of whaling and echoes of seals that bark in the protection of Point Lobos, are exceptional features along a harbourless coast.

In transverse profile the intensively cultivated valley bottoms are flanked by foothills on which as many as six species of deciduous oaks and the widespread, though intrusive, oat grass conjure up an appearance that was already observed as park-like by Captain Vancouver. The foothills are still devoted to animal husbandry, though they are increasingly likely to consist of milking cows intensively grazed on improved pasture. Seasonal aridity calls for careful fodder conservation and piped water. The foothill country is spring country, hot as well as cold, and has a variety of medicinal waters that the Old World Mediterranean would have exploited as spas. In more northerly valleys, the deciduous oaks yield to giant redwoods, especially on the seaward-facing slopes in the 'fog belt'. Intervening *sierra* and hill country is under *chaparral*, the New World counterpart of *maquis*. The evergreen oak mingles with the decid-uous; ceanothus, manzanita and buckthorn mix with sages, artemesias and lupins. The *chaparral*, usually 3–10 feet high, has a curiously patchy appearance, reflecting fire damage. Many tracts in the *sierra* are preserved as national parks for recreational purposes.

In latitudinal succession, specialist crops are adjusted to local modifica-tion in climate. Napa Valley, north of San Francisco Bay, has vineyards;

to the south, Santa Clara Valley has plums. Almond orchards (in the 'Almond rock country') yield southwards to peaches and apricots, to orange groves (now being replaced by avocado pears) and eventually to the sensitive lemons of Santa Paula (the 'lemon capital of the world'). The fruitlands, worth more than $2,000 an acre, fertilized, sprayed, fumigated, seasonally invaded by immigrant labour and co-operatively dispatching their processed products to the national or international market, represent the ultimate discipline imposed by man upon these valleys.

California's two principal cities are located in the coastal zone. San Francisco lies at the only low-level break in the littoral ranges and thrives upon its maritime contact. Los Angeles, spread upon the only extended portion of the coastal plain, prospers despite its dissociation from the sea. San Francisco is among the group of urban centres that ring San Francisco Bay and climb up its adjacent hills. Across the bay, which is spanned by long bridges, lie Oakland, Berkeley (with its university) and Alameda (with its navy yard). Southwards, the city's suburbs extend through San Mateo and Palo Alto virtually to San José in the Santa Clara Valley. The present city sprang out of the Spanish settlement of Yerba Bueno and is the second-ranking urban agglomeration on the Pacific coast. It processes the products of its back country (e.g. sugar beet, petroleum, animal products) and of its Pacific trading area (e.g. sugar cane, copra, newsprint, ores). Natural gas is a primary source of energy and San Francisco is integrated into the grid of the south-west. The Golden Gate which gives access to its harbour is the gateway to the Pacific. Much history has been crowded into the life of San Francisco in a very short time and this, no less than physical setting, ranging from tidal slough to Tamalpaian height, gives to the city a personality rivalled by few American towns.

Los Angeles lies between the mountains and the sea. Landwards, the San Gabriel and San Bernardino ranges rise to altitudes in excess of 10,000 feet. Seawards, there is a succession of broad beaches from Santa Ana to Santa Monica. The Los Angeles *pueblo* has become a polynucleal conurbation rather than a city, its predominantly single-unit dwellings bound together by millions of vehicles which operate on one of America's most complex road networks. Its growth has been twentieth century, erratic and spectacular, pushing south to San Diego to form a Pacific seaboard megalopolis. Its growth has been dependent upon a number of specialist activities, with a relatively high degree of dependence upon outside sources of supply or outsidemarkets. Five components may be seen in its economic structure. High-grade, high-value farm products gave Los Angeles County first place in all America until 1950, but farmland is being rapidly overrun by buildings and roads. Secondly, the motion-picture industry gave a generation of prosperity. Thirdly, the aircraft industry, entering during the war years, shows a remarkably dispersed distribution in the urban district. (The city of San Diego to the south also has aircraft industries.) Automobile manufacture has followed. Integrated steel mills in Los Angeles and at Fontana (combining Utah coal and Eagle Mountain ore), aluminium and rubber are dependent upon imported raw materials, imported water and were formerly dependent upon imported fuels. Distance from markets was an initial tax on sales' costs: today, the market is local. Deep-drilling for Californian oils over the last generation has changed the industrial background to Los Angeles. Oil occurs within the conurbation itself and in Santa Barbara, Ventura and Santa Maria along the coast; while so-called 'tideland oil' in Pacific waters is a reservoir for the future. Los Angeles thrives upon overcoming obstacles. Even its port at Long Beach, according to one method of reading statistics the largest on the Pacific coast, is an artificial creation. Yet it has acquired a fishing function larger than that of any other Californian harbour. Atmospheric pollution, related to local climate, local relief and automobile exhaust, is an outstanding challenge in a state with technical research institutes which rival any in America. Although the city's horizons may be blurred by smog, the world's largest telescope on nearby Mount Palomar owes its setting in part to the proverbially clear skies.

Beyond the coast ranges is the Central Valley (Plate VII), a structural depression 450 miles long and 50 miles wide, with the Sacramento River in the north and the San Joaquin in the south. Its low-lying floor has a deep covering of detrital material brought down by the many surrounding 'creeks'. The courses of the Sacramento and San Joaquin lie nearer to the coast ranges than to the Sierra Nevada: they unite in the saltwater delta behind the breach of the Golden Gate. The Sierra Nevadan tributaries have deposited impressive alluvial fans, which have been primary settlement controls. The valley grows progressively drier from north (20 inches) to south (less than 10 inches), and the effectiveness of the markedly seasonal rainfall diminishes sharply to the south of Fresno. Here there are dry lake beds, tracts of sand dunes, intermittent streams and interior drainage basins which have only recently been transformed into a chequer board of cultivated fields.

The transformation is a part of the large-scale management of water resources in the Central Valley. The 1870 report of the Commissioner of

Agriculture declared that 'the Great Valley is admirably adapted to agriculture on the grandest scale'. But the quantitative and qualitative (cf. chemical unsuitability) distribution of water resources was very uneven. In the late 1920's a unitary plan was formulated for the entire valley: 10 years later it was initiated. Flood control, generation of electrical energy and provision of navigational routes have run side by side in this multi-purpose project. In principle, the surplus waters of the Sacramento, with the Shasta Dam as its major control, are pumped beyond the delta area along the San Joaquin Valley to Mendota; while the surplus of the San Joaquin, controlled by the Friant Dam, is diverted into the Friant-Kern Canal for consumption by the arid lands around Bakersfield. The scheme has been a boon for all except the delta lands, where diminished water levels have prompted saltwater intrusion of some areas drained with the aid of oriental labour in the late nineteenth century.

In the human evolution of the valley, several distinct phases may be recognized. It was ranching country in the Spanish period and a vigorous grazing economy still persists around its margins today, where the 'cow counties' support one animal unit for every 2 acres of oakwood, or every 18 acres of pinewood. The discovery of gold in the stream gravels, below the 'Mother Lode' of the northern Sierra Nevada, gave rise to a multitude of mining settlements. Some persist, such as Grass Valley, Gold Run and Auburn on the main highway from Sacramento (the state capital) to Reno. The names of early farm pioneers are inscribed on the valley floor. Sutter symbolizes the third phase of large-scale wheat farming. He introduced commercial wheat farming, employed the first mechanical reapers, chartered clipper ships to bear the cargo to Europe, and left his name on the land. Today, specialized fruit and vegetable farming is conducted side by side with mixed farming (Figure 47). In few parts of the world can such a variety of crops be seen in juxtaposition. Surface flooding, furrow irrigation and sprinkler bring to perfection asparagus and sugar beet, rice (sown and fertilized by aircraft), corn, grapes and cotton. Cotton is California's fourth field crop and occupies a third of the cultivated area of the San Joaquin Valley. Perfection, indeed, is the object. In one respect distance isolates positively. It frequently detaches California from eastern plagues and pests. Inspection for fruit and vegetables is fiercer on the Californian boundary than at most European frontiers.

Against such an agricultural background and given the large local market, large-scale processing of both animal and vegetable products has arisen. To old-established sun-drying (with prunes in Santa Clara Valley

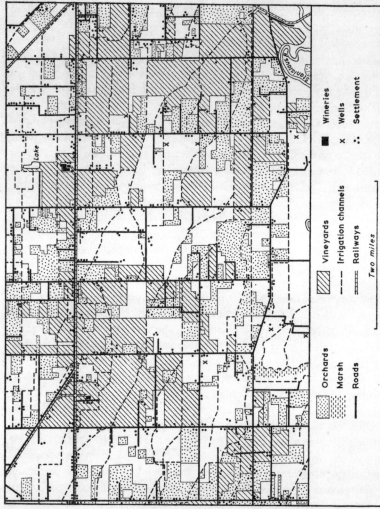

Figure 47 *Settlement and Land Use around Fresno (Cal.)*

The Central Valley of California reaches one of its peaks of agricultural intensity around Fresno. Farming is largely dependent upon irrigation, the essential features of which are shown on the map. Farms are small, settlement dispersed and orchards dominant (Source: *U.S.G.S.*, 1 : 24,000 Avena Quadrangle)

Orchards

Marsh

Roads

Vineyards

Irrigation channels

Railways

Wineries

Wells

Settlement

Two miles

Lake

and raisins around Fresno as outstanding examples) have been added canning, dehydration and deep-freezing. A 9-month growing period enables maximum economies to be made in processing plants. Wine is also widely produced, though there are no Californian counterparts of 'château' conditions. The processing industries have been much stimulated by the rise of cheaper thermo-electric energy resulting from petroleum and natural gas developments. The southern reaches of the Central Valley have petroleum deposits.

The valley floor retains little of its original vegetation. The plough, grazing beasts and fire have destroyed its perennial tuft and needle grasses. Much lumber has been removed from the surrounding foothills. Yet the Sierra has altered little and makes its own contribution to the fable of California. Its sky-high altitudes are being protected as national parks.

Imperial Valley, at the head of the Gulf of California, enlivened by the Coloradan waters of the All-American Canal, is the great home of lettuce cultivation. It is also sensitive to its international location, for so-called 'stoop labour' flows in accross the border from Baja California to harvest this crop. As many as 450,000 braceros have legally entered America seasonally from Mexico. Many more 'wetbacks' do so illegally. They are virtually American gypsies, their dreary encampments the object of improvement by the Farmer's Home Administration. Farmers prefer where possible to mechanise them and their problems out of existence.

THE INTERMONTANE HINTERLAND

From the human standpoint, the intermontane hinterland suffers most of the negative features and benefits from few of the positive features of the Pacific foreland. It is distinguished by verticality in topography and aridity in climate. Both have isolating effects. The two most extensive physiographic regions are the Great Basin and the Colorado plateau. They are omnibus terms, for the Great Basin (so-called by Frémont) is a complex of many basins broken by many ranges, while the Colorado plateau embraces and is intruded upon by many plateaux. The words 'basin' and 'plateau' conceal the immense vertical variations that intercept movement in a land which people go through rather than to. The Great Basin, essentially an area of inland drainage, is included principally within the states of Nevada and Utah: the plateaux are chiefly in northern Arizona and are tributary to the Colorado and Little Colorado rivers. Southern Arizona looks to the Gila River.

Landforms have been shaped under a semi-arid climate. They assume their most distinctive character in the south-west, where the horizontally bedded sedimentary rocks give rise to the broader flat-topped *mesas* and more irregularly profiled *buttes*. These *inselberg* forms may rise above the seemingly monotonous surface to heights of hundreds of feet. The plateau is, however, much more broken by the deeply entrenched *canyons* of the Colorado and, to a lesser extent, Gila systems. The Colorado *canyon* displays cliffs several thousands of feet in height. Rivers are therefore not only infrequent and inconstant in flow, they are often inaccessible. They isolate rather than unify. Arizona has a theatrical landscape. To the 'inverted mountain ranges' of its canyons are added volcanic features. There are intrusive laccoliths, ash-coned volcanic peaks active in A.D. 1066 (as in Sunset Crater National Monument), and twisted lava flows. The petrified forest symbolizes an earlier catastrophe in geological history. There is even a meteor crater. And the entire area offers one of the most highly coloured landscapes in the continent. Salmon-pink, terra cotta, tan, sienna, buff and tawny yellow cast blue and violet shadows in the Painted Desert. Most of Arizona supports some vegetation. There are scattered stands of conifers, with the ponderosa pine growing up to the 9,000-foot limit; creosote bush and burroweed provide a scrub vegetation and there are intermittent yuccas and aloes. Along the Mexican border there are patches of true desert and protected stands of organ-pipe cacti.

The contemporary traveller passes through this land in fair comfort. In summer he may journey by night and halt by day at the clustered motels that mark today's caravan stops. As he leaves New Mexico he will have been asked 'Are you prepared for 700 miles of desert?', slung a water bag over his radiator, closed his windows, turned on his air conditioning and be appreciating to the full a windscreen tinted against the glare of a burning-glass sun. In winter, Route 66 is a favoured road west because it offers the only snow-free passes. The washes, gulches and creeks across which the barbed-wire-bounded highway passes may be no more than strings of muddy pools pecked as blue lines on the map; but they will be raging torrents in less than an hour with the coming of the rain. The interior of Arizona may not receive more than 5–10 inches of rain, but a quarter or a third of the annual total may fall in an hour. Violent erosion ensues. Independently of this, the momentary surfeit of water in a land of annual deficiency is a distracting experience.

Snakes, scorpions and poisonous spiders remain, but in one respect this is a land of diminished human hazard. The native Indian has been civilized. Agriculturally experienced tribes, growing corn, gourds, beans, weaving, making pottery and baskets, were established in the south-west in pre-Columbian times. Their several-storeyed, brick-built dwellings and sandstone cliff-dwellings—scarcely less impressive than the fabled halls of Montezuma—may be seen today. The names of the Hopi, the Navajo and the Apache are still on the land. The Apache was an especially persistent outlaw, only entering reservations in the 1880's.

Inhospitability also characterizes the Basin and Range Province. Chain after chain of ranges with 10,000-foot peaks separate arid valleys with sinks (e.g. Carson Sink), half-dry lakes (e.g. Mud Lake, Nevada) and dry lakes or bolsons with their terraced, salt-encrusted *playas* or beaches telling of a former wetter period. Salt-flats flash white by day and resemble crystal snowfields by night. They reach their greatest extent in the Great Salt Lake Desert, where the Bonneville Flats provide an almost billiard-table surface for high-speed driving and for the spirals of the 'dust devils'. Vegetation reacts to slight variations in precipitation and the chemical content of the soil. In Utah the Engelmann spruce takes precedence on the western mountain flanks: sagebrush, about a yard high, casts a grey mantle over the flatlands.

In view of its wasteland character, it is scarcely surprising that most of this intermontane hinterland is publicly rather than privately owned. The percentage of federally owned land is 87 in Nevada, 70 in Utah and 44 in Arizona. Such land is managed by a number of different authorities. For example, the Bureau of Land Management maintains the grazing districts, while the U.S. Forest Service administers the woodlands (of which about 90 per cent belong to the Public Domain). Sometimes, control by Washington is resented as absentee landlords are resented in other parts of the country.

The keynote of the intermontane area is one of extensive, low productivity land-use, with a scatter of isolated patches of highly intensive activity. As with the High Plains, this is primarily livestock country, with Nevada most dependent upon it. Livestock management varies widely. Density per acre is usually low, dropping to 200 acres per animal unit in the extreme south-west. Elaborate transhumance to high-altitude grazings is found in eastern Utah and parallels that in adjacent mountain states. Journeys of several hundred miles may be made, with the aid of train or truck. The Biblical pastoralism of the Navajo Indian picturesque in coloured blankets, with his flocks, is a far cry. Sheep raise problems. Their numbers tend to decline proportionately to those of cattle and they are pushed to the marginal lands. They are more destructive to vegetation than cattle and are also less productive per unit of labour. Wool finds its principal market in the north-east: meat is marketed principally in the west. Indeed, this is meat-deficient country.

Intensive farming calls for careful water control. The early and intricate management of snow-fed streams at the foot of the Wasatch Range in Utah (Plate VIII) contrasts with the large-scale dams along the course of the Gila River. Near to the confluence of the Gila and the Colorado, there are major constructions such as the Imperial Dam, which provide water for the carrots, lettuce, melons and sweet potatoes of the Yuma area. The early Mormons were active disseminators of agricultural ideas—as seen in journals such as *The Farmer's Guide* (from the 1860's) and *The Utah Pomologist* (from the 1870's). Grapes, apricots and cherries thrive on slopes above the hayfields and below the sagebrush grazings. The contemporary counterparts of Mormon initiative are the newer crops that have entered Arizona. Cotton has made its mark and Arizona has had especial success with the Upland and American-Egyptian types. Sorghums have also been increasingly introduced, while throughout the irrigated fodder lands of the intermontane area, alfalfa predominates. It is harvested mechanically, and frequently dehydrated and converted into meal and concentrates.

Other than agriculture, the most intensive activity is associated with the extraction and conversion of minerals (Figure 48). Precious metals drew their share of speculators to these isolated lands. Nevada was historically 'The Silver State' with its Silver City, and silver dollars are still minted for its gambling saloons. Today the line of 'The Comstock Lode', basic to Nevada's reputation, is more haunted by surrealist ghost towns than fruitful with productive mines. Silver derives mostly as a by-product. Copper is more important. Bingham (Utah) with its celebrated man-made canyon, Yerington (Nevada) and Ray (Arizona) are mines among a score of others that enable the south-west to produce about two thirds of America's copper output. Lead, zinc, manganese, molybdenum and uranium are also mined. Iron ore is sought for the steel plants and Eagle Mountain mine in the Colorado desert is the chief producer. The West's largest steel plant at Geneva (Utah), barely 20 years old, rises out of the agricultural oasis of the Salt Lake area and enjoys a strategic isolation rare for a steel complex. Alkali deposits are widespread, e.g. Wendover, Utah (potassium salts) and Searl's Lake, Mohave (borax). Oil occurs in the intermontane lands, e.g. Uinta field (Utah), Ely field (Nevada), and low-grade coal is found in eastern Utah. Considerable hydro-electric energy derives from water conservation projects; while high-pressure gas lines, insignificant in appearance but potent in effect, traverse the countryside.

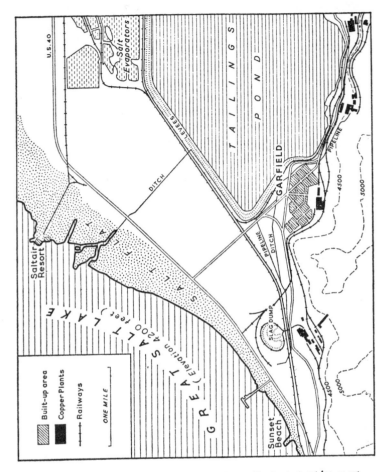

Figure 48 *A Mineral Landscape from Garfield (Utah)*

The salt workings of the Great Salt Lake and the components of a copper smeltery are seen here in juxtaposition. They offer an industrial contrast to the agricultural oases and to the steel industry of Provo to the south (Source: U.S.G.S., 1:24,000, Garfield and adjacent Quadrangles)

Most of the smaller settlements have specialist functions, from the provision of strategic materials to the serving of tourist needs, and visually they are often lost in the wilderness. The bigger cities have an almost too splendid isolation. Salt Lake City, founded in 1847, is oldest and unique among them; Phoenix is the state capital of Arizona. The liberties of Nevada in such resorts as Reno or Las Vegas seem licence beside the restraints of Utah, and they point the significant geographical effects of state boundaries. There is space for all forms of human expression—even for the explosion of atomic weapons.

THE PACIFIC NORTH-WEST AND 'THE INLAND EMPIRE'

In the North-West the features of the Cordilleran system are more concentrated and, although the plateau character is retained in southern Oregon and Idaho, it is changed in Washington. Three fundamental facts of physical evolution add distinction. First, although much of the Cordilleran country has been subject to the effects of volcanicity, Washington and Oregon can claim its most extensive tracts of basaltic lavas. Older lava rocks, inter-bedded with sedimentaries eroded by melt-waters and overlain with loessal soil, form the basis of the Waterville plateau, the Palouse Hills and their so-called channelled 'scablands'. Along the Snake River, Quaternary basalts and younger cones give rise to 'the Craters of the Moon'. Secondly, the intermontane area of the North-West has been more fundamentally affected by ice than the South-West. Thirdly, the North-West has a unifying river system. The Columbia River, gathering together the waters of south-eastern British Columbia, is central to Washington state: its principal tributary, the Snake, is the focus of interior Oregon and of Idaho (Figure 49).

Latitudinal location brings changes in the climatic régime. On the western flanks of the Cascades, there is abundant precipitation (80–100 inches p.a.), while in the troughs of the Cowlitz-Willamette valleys it may be a quarter of this. To the east, the aridity of the Basin and Range Province is continued northwards (10–15 inches p.a.), though the torrid summer heat is relaxed and precipitation effectiveness is greater. Snowfall is of great importance as a reservoir for summer water supply. This has favoured a heavier and more continuous woodland cover, with frequent climax conditions on westward-facing slopes. Good soils predominate, with abundant brown forest soils and broad expanses of chernozem and chestnut soils in the interior. While many of the western states write off

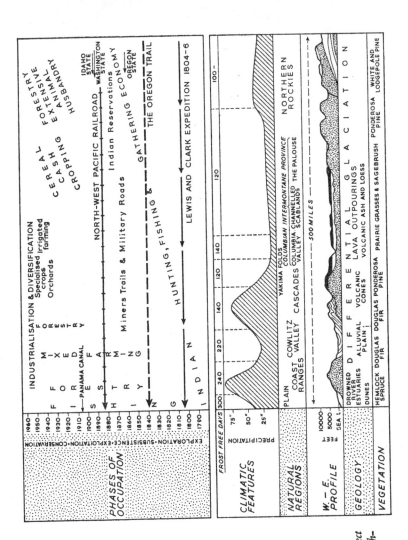

Figure 49 *A Transect through the Pacific North-West*

extensive territories as badlands, the Pacific North-West is more concerned with estimating the good lands—lands good for cultivation, for grazing and for forestry, and the good waters—waters good for the rehabilitation of fisheries.

Timber predominates in the north-western scene. Softwoods take precedence, the Douglas fir on the lower coast and Cascades ranges yielding northwards to spruce-hemlock woods with lodgepole pine at higher elevations. Inland, the ponderosa pine takes over the drier sites. There are oak groves, especially in Oregon, and larch stands.

Almost half of the North-West is too dry or too rough for cultivation and is used as range land. Beyond the timber stands are the perennial bunch grasses, sagebrush (which invades overgrazed grassland) and salt scrub: at higher altitudes are mountain grasses: of more limited occurrence are hardwood groves and pastures. The eastern interior, together with adjacent Montana and Wyoming, has evolved some of the most elaborate as well as large-scale stock migrations found in the world.

Farmland is principally a response to water availability and, as would be expected in a countryside of such diverse relief and bedrock, its use displays marked regional variations. The coastal valley areas have old-established, mixed farms, with city-supply types of product. Fodder crops are easily grown under irrigation where necessary in the Willamette Valley, and many tracts have year-long grazing. Dairying, with butter and cheese as end-products, is also a specialization of interior Idaho. Wheat and small grains, both fall-sown and spring-sown, are a feature of the interior plateaux around Spokane. Sugar beet, onions, and above all potatoes, have been developed in Idaho. The tree-fruit crops of Washington and Oregon are, however, the most spectacular. In Oregon stone fruits predominate: in Washington, apples and pears are intensively cultivated in the spacious irrigated valleys of Yakima, Wenatchee (with 50,000 acres of apples alone) and Medfoot on the eastern side of the Cascades. Beyond the sown lands are the grazed lands, commonly held as large stock ranches (500–5,000 acres), which are as likely to consist of several as of one unit. In agricultural resource, the Pacific North-West virtually bursts at the seams. After a brief period of ignorant rather than irresponsible exploitation, the scientific management of woodland and pastureland and the maintenance of soil fertility are being assiduously practised.

Principally because of isolation, large-scale exploitation of the resources of the Pacific North-West was retarded. Overcutting, overstocking, overcropping and overfishing only affected limited areas before restraints were introduced. The improved management of the resource base is inseparable from the changed status forced upon the Pacific North-West during the second world war and sustained there since. The rise of strategic industries has called for the fullest exploitation of its power and water supplies. Industry and agriculture in the broadest sense have been jointly interested in the co-ordinated development of the Columbia-Snake system.

The accent has been on the Columbia River, which maintains a large volume of water with a relatively low silt burden, has steep gradients and good sites for damming. Agriculturally speaking, the river, having its maximum run-off in spring, can offer assistance at the critical growing period. It is also central to extensive tracts of semi-arid land. The construction of the Grand Coulee Dam with Lake Roosevelt was the first milestone. The Bonneville Dam and the Chief Joseph Dam at Bridgeport are other major enterprises, but on the Columbia-Snake system in its entirety only a few dozen out of a full hundred large-scale sites have been built up. The series of dams also play an important part in flood control. Integration of power resources is extended through to Idaho, Montana and Utah by the North-West Power Pool, the grid of which spans two time zones. The North-West states claim rather more than a third of the U.S.A. hydro-electric power potential. This is the more important because of the regional deficiency of combustible fuels.

The first industries to use water power were those associated with forestry. Forest operations are increasingly mechanized, from tree-topping equipment to multi-wheel logging trucks, from caravaned logging sites to portable sawmills. The Douglas fir is the principal tree sought by lumber mills. Plywood, pulp and paper processing, subtracting weight from raw materials and adding value through manufacture, have grown rapidly. Overcutting in the coast ranges and valleys has resulted in a shift to the Cascade slopes. A fifth of America's pulp, a tenth of its paper and most of its softwood plywood derive from Washington and Oregon today. The North-West Pacific woodlands are a long way from the national markets, but shipment via Panama (14 per cent today), the railroads and more recently by truck has confirmed them as the primary source of constructional timber.

The cheap power potential already attracted several groups of industries to the Puget Sound before the second world war. Their concentration was increased for strategic reasons during the war and their subsequent multiplication has left the Pacific North-West on the edge of a power crisis. Nearest to hand are the underdeveloped resources of the Fraser River. British Columbian administrators are reluctant for these to be developed for export, but have agreed to co-operate on the upper reaches of the Columbia River. Trans-Cordilleran pipe-lines bring in natural gas and oil from Alberta, while Californian oils also help to offset power deficiency.

The principal consumers of power are the electro-metallurgical industries. Aluminium smelteries, drawing their raw materials from Jamaica and the Guianas and overland by the long rail haul from Arkansas, are located in Mead, Tacoma, Vancouver, Troutdale, Longview and Wenatchee. Steel foundries and rolling mills, using imported ingots and local scrap, are established in Seattle, Tacoma and Portland—this last city has a 35-foot river channel. There are refineries (Idaho is America's leading silver producer and Montana the chief copper producer), ferro-alloy plants, electro-chemical industries and atomic-energy installations at Hanford and Arco (Idaho). The products of these primary industries are mostly absorbed in the shipyards and mechanical workshops around Puget Sound and the lower Columbia Valley. There are also aircraft works, which antedate the aluminium industry. By comparison, the hundreds of food-processing plants, dealing with dairy products, fish, fruit and vegetables, are very modest consumers of energy.

The Pacific North-West has always looked to the land for its wealth. A succession of primary staples have sustained its down-to-earth economy. The fur trade, sporadic precious-metal mining, large-scale ranching, bonanza wheat and heavy lumbering have all left their legacies in daily life, but industry, though recent, is the centre of the current phase of development. Washington, Oregon and Idaho are an industrial problem area and isolation is a primary cause of their difficulties. An extractive economy with raw-material processing as the main industrial activity implies that, in general, production is of low value in proportion to bulk. Such goods are also comparatively limited in variety. There are relatively few secondary or tertiary manufacturing industries and the Pacific North-West is sensitive to trade disturbances. In addition, the short supply of skilled labour results in high wages and considerable mobility. While detachment of the area has accounted for the late industrial start, it takes a continuing toll in the freight costs of finished goods. Visually, the Seattle-Tacoma district presents an impressive industrial and urban panorama, but it is around the Puget Sound that most of the score of manufacturing counties are situated. The status of industry in the North-West is given new perspectives in view of the realization that it is the seat of only 3 per cent of America's manufacturing activity.

But if the North-West is remote from the main American market, it is on the threshold of the Pacific. This fact asserts itself most strongly in the ports of the Puget Sound. Seattle, for example, is headquarters of 80 steamship lines and deals with 4,000 cargo vessels a year. In 'Fisherman's Terminal', on appropriately named Salmon Bay, Seattle can harbour 1,000 vessels in the North-West's largest fishing port. It has become the more important with the rise of Alaska, for which it is a point of departure strategically, commercially and even academically.

THE FORTY-NINTH STATE

In Alaska the physical features and human problems of the Cordilleran province are transferred to higher latitudes and are simultaneously exaggerated. Alaska, with half a million square miles, is the largest American state; with about 100,000 inhabitants, it is the least populous. There is a steady subtraction of opportunity northwards through British Columbia and the Yukon into Alaska. But although the assets and liabilities of the forty-ninth state are continuously in mind, they are not estimated in quite the same way as those of any other parts of North America. Strategy as well as economy enters its balance sheet.

The form of the land increases the sense of detachment. The arcuate wall of the Alaskan Range, fiorded in the Panhandle, loftiest in the central Mount Elias Range, extending north-westwards in the peninsula, isolates the interior from the Pacific more absolutely than anywhere along the entire coast. It harbours some of America's largest glaciers, e.g. Malaspina, an ice cap probably 2,000 feet thick. In the Panhandle, there are blunt-headed fiords of more than Norwegian magnitude, etched by wave-cut terraces which are still in process of rapid emergence. Tide-water glaciers, e.g. Columbia, Fairweather, and Guyot, with ice cliffs 100–200 feet high, calve into salt sea. Westwards is the arc of the Aleutian islands, stepping-stones to Siberia and Japan, and the *Ultima Thule* of America.

Access from the warm-water coast, with its dense fogs and heavy clouds, its gales and blizzards, is restricted to a few valley ways, of which the Sutsina at the head of the remarkable Cook Inlet, and the Copper-mine in the shadow of Mount Wrangell Range, are principal. The interior focuses on the Yukon River with its tributary valleys, the entire system of which has been excavated from the massive sedimentary infill between the complex of coast ranges and the interior Brooks Range. The interior lowlands are oriented to the shore of the Bering Strait, with their

R

extended coastal and estuarine plains. North of the naked Brooks Range is the desolate Arctic slope. It is true tundra, retaining snow until mid-summer, when it is converted into morass and mosquito land.

Almost the whole of Alaska suffers a surfeit of swampland. On the western flanks this results from heavy rain and high humidity; inland it is explained by immature drainage: seasonally, it results from snow and ice-melt and the surface thawing of permanently frozen subsoil. In the vocabulary of American geographers Alaska is the classical land of permafrost. Its thaw-lakes, polygonal ground patterns, pingoes, mud-flows and stone stripes have been investigated in their practical as well as their academic contexts. For satisfactory roadbuilding and engineering, airfield and landing-strip construction are all dependent upon the correct calculation of the permafrost impact. From the façades and floors of the city of Nome, which have tilted as the warm foundations of buildings have settled, to the continual discovery of mastodon and mammoth remains refrigerated since the Pleistocene, permafrost offers its testimony for the curious.

The land of ice is also a land of fire—a part of the Pacific's 'fiery ring'. North of the Isthmus of Tehuantepec, only Mexico has such active volcanism as Alaska. From the peninsula to Attu island over forty volcanoes are known to have been active during the last two centuries. The Valley of Ten Thousand Smokes is but one place-name which testifies to this activity. Lava spreads and ash deserts are complemented by hot springs. The interior plateaux have a dozen sites at which hot spring-water flows abundantly, though Iceland's example in their employment for central heating and greenhouse cultivation has not yet been followed. Earthquakes add to the natural hazards of avalanche and rock-fall in this precipitous land.

Indian and Eskimo are native to Alaska and their place-names mingle with those of Russian traders and settlers, men of fortune from the four corners of the world, American administrators and heroes. Shishaldin volcano on Unimak island is a nice juxtaposition of two place-name elements. Sitka, the old Russian capital, replaced by Juneau in the Panhandle area, has sent its name round the world with that of its local spruce. Gold, which drew a cosmopolitan company, imprinted its own picturesque toponymy—Hard Luck, Last Chance, Helpmejack.

Among the resources of Alaska, the fauna and flora have made the most continuous contributions. Both Eskimo and Indian had a hunting economy—the former directed to seal, walrus, whale and waterbird: the latter, to caribou: both to the musk-ox. Intrusive Russian and American hunters and fur traders preyed upon their preserves, in particular exhausting the sealing grounds. Nowhere has conservation shown more rewarding results than in the protection of pelagic seals in the Pribilov Islands. In the 1890's, attempts were made to naturalize Siberian reindeer in the interior; but although they thrived initially, numbers have dwindled to about 50,000. Overstocking of grazing lands and the effects of fire upon pastures are two possible causes. Reindeer have also competed for pasturage with musk-ox. Fisheries are the principal source of contemporary wealth. The five salmon species (Sockeye, Chinook, Coho, Humpback and Chum) are the chief catch. They call in the great 'salmon air-lift' of seasonal labour to man the floating canneries and mainland canneries that operate from May to September, and the cold-storage warehouses and curing plants that have multiplied rapidly over the last generation. Trespass and dispute has been succeeded by management and control. The International Pacific Halibut Commission, for example, arranges seasonal agreements with Canada. The less valuable herring is reduced for oil and meal.

The most accessible resource of the land is timber. Altitude and climate limit the extent of the timber cover and restrict it primarily to the Pacific coastal areas. Three quarters of Alaska's timber is situated within a few miles of salt water and the best stands are in the Panhandle, where the timber line rises to 2,000 feet. Hemlock (basis of a paper and pulp industry), spruce, red cedar and Alaska cedar predominate. Forestry remains an enterprise with much promise.

Forest land complements farmland, but both animal and arable husbandry are small in scale. Soil experts have estimated that Alaska has a cultivatable area as extensive as the states of Illinois or Iowa, but it is land that has a growing season of 100 days or less, and summer frost risks are inescapable. Stock rearing is more suitable than arable farming, but 7 months of indoor feeding may be required. The Matanuska Valley contains some of America's last pioneers and also supports Alaska's chief experimental farm. Isolation discourages production except for the local market. It is both simpler and cheaper to import foodstuffs.

Most Alaskan settlement is urban rather than rural and has a transient character. Towns, such as Juneau in the Panhandle, Anchorage on the Cook Inlet, Fairbanks on the Yukon River and Nome on the Seward peninsula, have a shifting population. Fishing and forest communities consist only in part of permanently resident Alaskans. An army of technicians and research workers spreads itself thinly over the land, its personnel moving in and out on short-period contract. Members of the

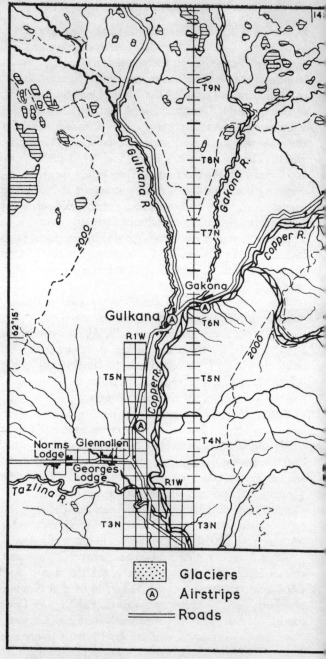

Figure 50 *Topography and Settlement around Gulkana (Alaska)*
Gulkana is a meeting place of Alaskan valleyways to which
twentieth-century communications have brought a lively touris

Sinona
Lodge

Chistochina

2000

3000

4000

5000

6000

Sandford R.

4000

5000

6000

Mt.
Wrangell

Lakes and rivers
Buildings
10 miles

ndustry. Permanent ice sheets look down upon it and ice edge
phenomena dominate the landscape features (Source: *U.S.G.S.*,
1 : 250,000, Gulkana sheet)

services come in large numbers to the many air force and naval bases. Alaska is a growing recreation land for Pacific business men, with woods rich in game and inland waters full of fish. The Alaska Highway also brings in a summer flood of tourists. A ticket to Bering Strait may be bought at any tourist agency, while it is not unknown for an enterprising Eskimo to go as far afield as Hawaii for a holiday.

Most transient of all have been the settlements associated with former gold workings. Late nineteenth-century gold-rush towns such as Skagway and Dyea yielded to encampments such as Iditarod, which has accommodated as many as 5,000 people. Bases such as Nome housed summer populations of 20,000 sixty years ago. The pick-and-shovel days are over, but gold is still produced. Nome and Juneau have placer workings, though the large-scale mechanical removal of the frozen overburden adds greatly to production costs.

Extensive areas of Alaska await survey, but it is evident that there are abundant and widely scattered mineral deposits. Tertiary coal measures occur in the Matanuska Valley and elsewhere along the narrow coastal lowlands. Copper occurs widely though mining has been sporadic, as in the Kenikott mines of the Copper River. Platinum occurs near Good News in the south. Above all, there is oil on Alaska's North Slope. In general these deposits suffer a double inaccessibility—distance from national markets and isolation within Alaska itself. The same applies to water power, the most obvious sites for which occur along the Yukon River. It needs an 800-mile pipeline to bring North Slope oil to the Gulf of Alaska.

The challenge of isolation combated partly by express coastal steamers based upon Puget Sound, eased locally by cumbrous railway lines, aided uncertainly by river boat, is met by a growing network of roads and airways. The Gulkana quadrangle (Figure 50) indicates the impact of these new services. Highways, with hunting lodges and fishing cabins, penetrate from the coast to interior Fairbanks, airstrips serve bush aircraft, while regular flight paths traverse the country to link its main towns with Seattle.

For outsiders, Alaska has two faces. One is drawn by adventurous romantics such as Jack London and Rex Beach, the other is portrayed by imaginative enthusiasts such as the first Alaskan Director of the U.S. Geologic Survey. 'Had the Pilgrim Fathers settled in Sitka instead of Plymouth,' he wrote, 'they would have found a milder climate, better soil and timber, more game, furs and fish.' This is the legend of the land of opportunity. The forty-ninth state remains an appendage, with the problem of absentee landlords and of dependence upon the federal government. It would require a nation, not a state, to create the neo-Scandinavia in Alaska which optimists still forecast for it.

THE CLIMAX OF THE CONTINENT

Americans are still going west and the seaboard states of the Pacific are their journey's end. The means of easy access are available, but human mobility is not necessarily matched by commodity mobility. This is reflected in the refining of Pacific coast raw materials before overland movement to a continental market; the establishment along Pacific shores of the branch plants of 'Atlantic America' and by the award of an impressively large number of unclassified defence contracts to California.

'California the golden' continues to take precedence as the objective of migration, though the Pacific North-West exerts a growing attraction. The adventure of Alaska has a minority and short-term appeal. Population increase has had two primary effects. First, the existing unevenness in distribution has been exaggerated by the sharp increase. Simultaneously, it has exaggerated natural contrasts at least as much as it has softened them. Secondly, while population multiplication is impressive, the material wealth which is its accompaniment is even more so. California (pop. 1960, 15,717,204) has become one of the country's greatest markets. Its rise is an aspect of America's return to the seaboard. Along the Pacific coast is seen much of the climax of the continent and of its development.

SELECTED READING

Standard American texts from the Pacific slope are

O. W. Freeman and H. H. Martin (ed.), *The Pacific North-West* (New York, 1954)

C. M. Ziarer (ed.), *California and the South-West* (New York, 1956)

Southern California was the object of a special publication of the *A.A.A.G.*,

W. L. Thomas (ed.), *Man, Time and Space in Southern California*, 49, 3.2 (1959)

See also: R. U. Cooke and I. G. Simmons, 'Some Recent Changes in California', *Tijdschrift voor Econ. en Soc. Geografie*, 1966, 232–2

The Univ. of California publishes a research series of geographiical monographs,

Isaac Steven, 'Practical Geographer of the early North-West', *G.R.* 45 (1955), 542–58—introduces Washington and Oregon historically

R. M. Highsmith, *Atlas of the Pacific North-West* (Corvallis, 1958)—measures the development of the same area

The native picture is filled out by

E. E. Dale, *The Indians of the South-West* (Norman, Oklahoma, 1949)

and a study of the technique of land settlement in three Utah and three west Canadian settlements is given in

L. Nelson, *The Mormon Village* (Salt Lake City, 1952)
D. W. Meineg, 'The Mormon Culture Region', *A.A.A.G.*, 55 (1965), 191–220

An historical monograph of continuing interest is

J. O. M. Broek, *The Santa Clara Valley, California* (Utrecht, 1932)

Two detailed crop studies are

D. C. Large, 'Cotton in the San Joaquin Valley', *G.R.* 47 (1957), 365–80
R. N. Young and P. C. Griffin, 'Recent Land Use Changes in the San Francisco Bay Area', *G.R.* 47 (1957), 396–405

See also

J. L. Parson, 'Californian Manufacturing', *G.R.* 39 (1949), 229–41
R. U. Cooke and I. G. Simmons, 'Some recent changes in California', *Tijdschrift voor econ. en soc. Geografie* (1966), 232–42

and on aspects of Alaska

Alaska Science Volume, issued by the Arctic Institute of North America (Montreal, 1952)
H. Williams, *Landscapes of Alaska* (Berkeley, 1958)
K. Stone, 'Populating Alaska', *G.R.* 49, 1 (1959), 76–94

Canada

9

Metropolitan Canada and Its Problem of Industrial Adjustment

The Variety of Environment—The Dichotomy of Peoples—The Tradition of Agriculture—Movement from the Land—The Basis of Energy—The Transport Network—Continuity of Leadership

THE VARIETY OF ENVIRONMENT

The St. Lawrence lowlands of Quebec and the Lakes Peninsula of Ontario were the hearts of Lower and Upper Canada respectively. They are also the background to 'Metropolitan Canada'—the most industrialized part of the nation, the seat of its government, the headquarters of its financial agencies, the site of its largest cities and of its principal port. The St. Lawrence lowlands, oriented northwards, narrow as the estuary broadens, expand about the relic of late-glacial Lake Peter and assume their greatest extent in the Montreal plain. The river, tidal to Quebec City, has a channel of 27 feet to Montreal. The canalized rapids of the St. Lawrence, with the Thousand Islands east of Kingston, reflect a pre-Cambrian interruption of the sedimentary lands. On the map the Lakes Peninsula of Ontario appears to have a more logical boundary along the axis of Lake Nipissing and the Ottawa River (the provincial boundary to the east of the Gatineau River). But the edge of the Shield lies much farther south. It follows fairly closely the zone of fractures stretching from the Bay of Quinte, by way of the Kawartha Lakes and the Trent River navigation to Parry Sound. To the south-west of this zone lies the sedimentary part of the Lakes Peninsula. Shales, limestones and sandstones yield to each other from north-east to south-west. The Niagara escarpment, 'the mountain' to those who live beside it, deeply dissected by the gorge of the Niagara River at the international boundary, is the most outstanding physical feature of the peninsula. It extends as a controlling feature into the Bruce peninsula and beyond to Manitoulin Island. Almost everywhere bedrock is covered by glacial deposits of varying thickness. Altitudes of more than 1,000 feet a.s.l. are reached in the end moraines of western Ontario. Extensive lacustrine deposits are encountered around the shores of Lake Erie and Lake Ontario especially. The lake shorelines are generally low-lying, much subject to erosion by wave action and characterized by sand-bar and lagoonal features. Lake Erie, 572 feet a.s.l., the most southerly of the Great Lakes, is relatively shallow, with broad sandy beaches: Lake Ontario, 246 feet a.s.l., is deep, and in size as extensive as Yorkshire.

From the human point of view, the most southern part of the Shield is closely integrated with metropolitan Canada. The Ottawa plains with their federal capital offer tracts of moderate fertility, the Muskoka Lake country is the summer playground of the southern cities (and also winter ski-ing country): Algonquin Park was one of the first provincial parks to be created in Canada. This area, however, is more commonly country of erosion than deposition, of podsols rather than brown forest soils, of the pine rather than the maple. While the Great Lakes peninsula is an extension of the Middle West into Ontario, this is an extension of the northlands into the Great Lakes peninsula. However, it adds its own particular contribution to the remarkable diversity of landscape and landform which makes the stretch of country from the Ottawa Valley to the St. Clair River one of the most inviting and rewarding of the settled areas of Canada.

The riverine lands of Quebec also have tributary territories, which are opened up by the rivers which drain from Laurentia and Appalachia. Downstream is the Saguenay and its fiord which gave access to the Lake St. John lowlands; midstream is the St. Maurice; upstream is the Gatineau. On the right bank the Richelieu River opens an overland route to Lake Champlain and the upper reaches of the Hudson, while the St. Francis and Chaudière penetrate into the Vosges-like Eastern Townships (Cantons de l'Est) along the Maine-Vermont border. These are lands of varied and variable opportunity, with restricted arable tracts along lake and river terrace, but extensive and accessible woodlands.

Diversity is also a characteristic of the climate of metropolitan Canada. East-coast influences exaggerate the raw winter cold of the St. Lawrence lowlands: the Lakes peninsula, lying in the same latitudes as Mediterranean France, has a modified continental climate. Metropolitan Canada is influenced by indraughts of warm Gulf air and cold polar air which give wide variations within the seasonal round. The interrelation of land and water make for marked climatic instabilities; while sudden changes of temperature may also occur. In May, day temperatures in the 70's may be succeeded by night frosts: in January, temperatures well below zero may be followed within 36 hours by a thaw. Continental rainfall in summer takes the form of thunderstorms; while a high humidity exaggerates summer heat and winter cold. The rhythm of daylight and darkness differs from that of Great Britain. Such temperature ranges and changes affect the plant community: British farmers and gardeners were frustrated in their attempts to bring field and garden practices from the 'Old Country' to Upper Canada. Permanent grass produces a very indifferent sward in such contrasts of heat and cold, drought and deluge: there is no mass of meadow flowers: lawns are less resistant and must be strengthened with sown seed: the short, uncertain spring upsets the orderly sequence of flowering from February snowdrops to May tulips, and when it does not kill them it crowds them all together. Quickset hedgerows fail: roses must be trussed up against winter frost. But native vines run riot; while the markets of such a city as Hamilton have a cornucopian display of fruits and vegetables in August. The warm summers and relatively long growing season bring the plains of Montreal virtually into the 'Corn Country'. Growing seasons may vary widely in short distances, e.g. 215 days along the Erie shoreline and 190 days in the nearby western uplands.

THE DICHOTOMY OF PEOPLES

Variety in the landscapes of metropolitan Canada is matched by an essential dichotomy in its peoples. This asserts itself strongly despite, perhaps because of, the increasingly cosmopolitan character of Ontario. Historically, Lower Canada is French-speaking and Catholic; Upper Canada is English-speaking and non-Catholic. The heart of Upper Canada has resided in its Empire Loyalists; of Lower Canada in its clerics. The separatism of the latter is matched by the strong imperial spirit of the former. The unitary origin of French-speaking Canadians has been but little impaired by intrusive elements, though more than 250,000 immigrants have settled in Quebec since the war. Ontario has received much larger numbers. The *Canadiens* have been more resistant to the invasion of American capital and ideas than Ontarians. The submission of English-speaking Ontarians to American loan words and their militant rejection by the *Bon Parler Français* movement of Quebec present extreme examples of this.

The province of Quebec also shows a concentration of settlement which contrasts with its historic dispersal of energies. There is expansion of the area of French-Canadian settlement, but into contiguous territories. The invasion of the Eastern Townships is a virtual *fait accompli*; contemporarily, there is a steady expansion into the Ottawa river lands of Ontario and tracts along the St. Lawrence banks above Lachine. Scottish Presbyterian farmsteads around Ottawa now announce French-Canadian names on their wayside letter-boxes. This expansion is partly explained by the swifter growth of French-Canadian population. The birth rate is about 10 per cent greater than among English-speaking Ontarios; the proportion of young people is higher; even the percentage of women in relation to men tends to be greater. The pioneering spirit seems to be stronger in the French-Canadian (and it is a spirit encouraged by his Church); his endurance in the face of agricultural adversity seems to be more pronounced; he is generally prepared to work longer hours and to subsist on lower standards than his English-speaking counterpart. Inter-provincial rivalries incline to disrupt the seeming unity of metropolitan Canada, but the adherence and integration of the two parts is essential for the well-being of the country. Industrialization tends to exaggerate the differences between the ethnic groups.

THE TRADITION OF AGRICULTURE

By tradition, the peoples of Upper and Lower Canada are farmers, first adapting European crops and stocks: later adopting those of the New World. Early Quebec farm censuses give some indication of this, and successive travellers' accounts, such as the journal of Pehr Kalm from 1749 or Joseph Bouchette's *Topographical Description of Lower Canada* (1815), provide glimpses of persistent and changing elements in its rural economy. The settlement of Upper Canada is well documented in Colonial Office archives, while a wealth of journals tell of the introduction of seeds, stock, equipment and practices from south of the border (especially from progressive New York state). Quebec has its farm censuses and, in Cornelius Krieghoff, an artist of no mean repute to record the rudeness of its frontier scene. The shape of the farmstead is still frequently guided by the lines of the historic *arpente*—and a precise example is provided by the farm reporting to the *Association of Agriculture* (cf. also Figure 51). Root fence and snake fence can still be seen to tell the tale of timber clearance around Ontario farmsteads. 'Of banking, hedging and billhooking, they know nothing,' wrote William Cobbett of adjacent Ohio, 'but cutting

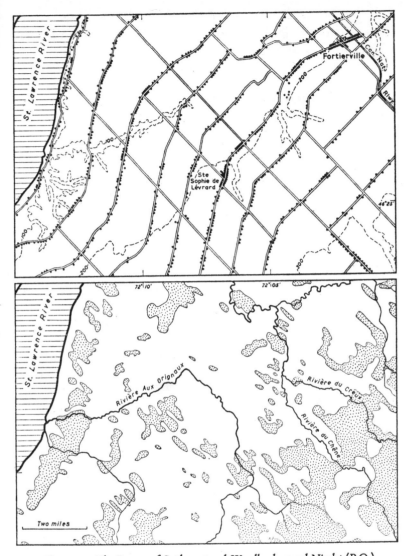

Figure 51 *The Pattern of Settlement and Woodland around Nicolet (P.Q.)*

Settlement along the St. Lawrence has its own distinctive pattern which is related to the *rangs* (or range lines) parallel to the course of the river. The *rangs*, usually decreasing in age of occupation from waterfront to interior, have given rise to their own groupings (e.g. co-operative dairies, schools, etc.) and even to their own idiomatic expressions (e.g. *vivre dans les rangs*). Their character is summarized in P. Deffontaines, 'Le rang, type de peuplement rural du Canada français', *Proc. 17th. Congress of I.G.U.*, Washington, 1952, (1957), 723–6 (Source: *Canadian G.S.*, 1 : 50,000, 1953)

down trees, they will do ten times as much as any man I have seen.' The attack on the woodlands of metropolitan Canada has been so intensive that less than a tenth of the area is forested today. Woodland clearance may explain a widespread lowering of the watertable with wells drying up in the Niagara limestone country and mill-ruins astride and beside stream beds which are seasonally empty. The old mill by the stream of Nellie Dean, famous in song, no longer has water!

Farming remains a valuable source of income for metropolitan Canada today and a primary basis of industry. It has been dominated by mixed husbandry since the decline of grain prices with the opening up of the Prairies. Dairying, with pigs and poultry subsidiary, is established upon a basis of mixed fodder crops. Indian corn and lucerne alternate with more familiar roots and cereals. Experiment characterizes rotations and field operations are heavily mechanized. Animal husbandry calls for heavy investment in farm buildings and 6 months' indoor feeding.

Side by side with Ontario's progressive mixed farming exist more localized specialist activities. Commercial fruit production prevails in the Niagara peninsula, around the Bay of Quinte and along the Erie coastlands. The Niagara lacustrine plain, 2 or 3 miles wide between the escarpment and the low Ontario shore, has peach, pear, apple and plum orchards (Figure 52). Late frosts, severe winter cold, disease and insect pests take intermittent toll of them and of the vines, small fruits, vegetables and gourds which are also commercially cultivated. Descendants of the original Loyalist farmers have to a large extent sold off their estates in small lots to family farmers of immigrant stock who are prepared to operate without costly hired labour. The Niagara peninsula experiences growing competition for land use—suburban development, factory expansion and highways swallowing up agriculturally precious land (Plate IX). The intensely cultivated north shore of Lake Erie adds tobacco as a cash crop to fields of tomatoes, peas, asparagus, sweet corn and pimentoes. Tobacco makes particularly heavy demands on labour during the harvest as mechanization is either expensive or ineffective—or both. An influx of pickers and tiers from the U.S.A. eases the situation, but university and high-school students are also prepared to lacerate their hands for the high wages offered. The Bay of Quinte is primarily apple country. Peatland cultivation of salad and vegetable crops, e.g. at Holland north of Toronto, extensive greenhouse cultivation (especially of flowers) and nursery gardening around the cities attempt to compete with American products which are speedily shipped from kindlier climes.

Marginal conditions in the tributary lands stand in contrast—the islands

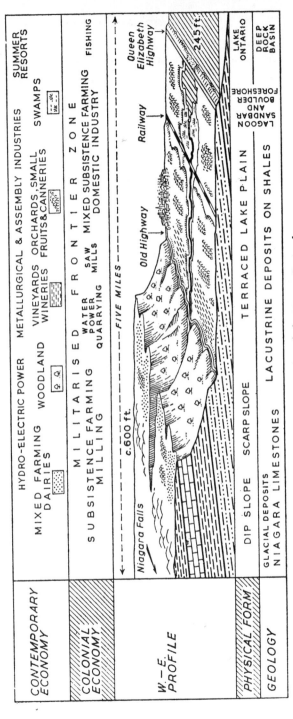

Figure 52 *A Transect through the Niagara Peninsula*

of settlement along the St. Maurice and around Lake St. John where lumbering (*bûcherage* to the *habitant*) takes increasing precedence over farming, and when field husbandry has often been regarded as something relegated to *les loisirs d'été*. This is the countryside which, curiously enough, gave birth to Canada's one classical novel—Louis Hémon's *Marie Chapdelaine*.

MOVEMENT FROM THE LAND

The evenly distributed farming settlements—with dispersed settlement predominant in Ontario, and nucleation around the church in Quebec —have had to face problems of soil exhaustion, soil erosion and an uneven distribution of rural services. Depopulation, which has characterized the Ontario countryside for a generation, has become increasingly pronounced in Quebec. The average Quebec family may be sufficient for the needs of the farm given older methods of cultivation, but the farm is insufficient for the full-grown family, given present standards of living. The establishment of single inheritance upon the death of the owner (with the *fils héritier* also inheriting a mortgage) side by side with migration from the home farm are attempts to maintain standards in French-Canadian farming. In Ontario productivity is pitched high and the province holds the lead in value of farm output despite rural depopulation. In Quebec the *habitant* retains a stronger attachment to the land and on the best farms displays an efficiency equal to that of the Ontario farmers.

Townward drift in Canada has been nowhere more pronounced than in Ontario, while the growth of the towns has been exaggerated by postwar immigration. The reduction of farm labour has been partly offset by mechanization. It is also partly reflected in the local retreat of the frontier of cultivation, though this may sometimes be from areas which would have been best left under their original forest cover. The drift to the towns, in some respects more necessary, has been generally more serious for the Quebecker than for the Ontarion. The rural migrant in Quebec is likely to be a less skilled and less adaptable worker than his Ontario counterpart. He will commonly come from a farm which is owned by his family, but will generally become a tenant in an urban home. The Ontario will probably move from a tenanted farm (more than a third are tenanted in Ontario), but will rapidly set about the purchase of his own home in a town. The type of town towards which he will move will probably differ considerably in the two provinces; though the towns of Ontario all have their French-speaking immigrants. Two thirds of the population of Ontario is located in five cities. The degree of concentration of population in Quebec Province is less. This is partly because of differences in the scale and degree of industrialization.

The growth of Canada's principal cities, Toronto and Montreal, is only partly the complement of rural depopulation. Both have attracted a large immigrant population. Montreal, founded 1641, a thousand miles from the open ocean and surrounded by a circle of tributary territory, has an international hinterland which extends into the U.S.A. (Figure 53). It

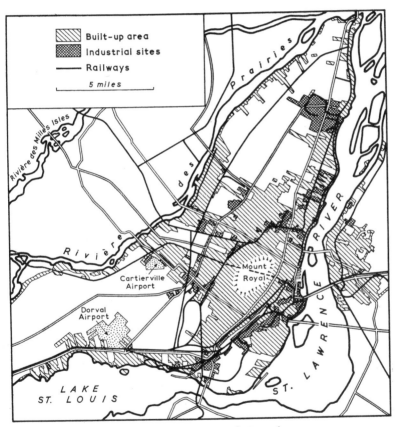

Figure 53 *The Setting of Montreal*

Montreal is concentrated essentially between the volcanic Mount Royal and the waterfront of the St. Lawrence, the river terraces of which provide guide lines in its street pattern. The city has outgrown its island setting, so that new industrial and residential sites are sought across its surrounding waters

is also located on a provincial frontier, at the meeting ground of French-
and English-speaking Canada. Montreal is a strong point binding the two
parts of metropolitan Canada together economically. Toronto, originally
old Fort York, has a lakeshore site and a superficial area reputed to be
larger than that of Paris. It is one with the lakeshore cities of the U.S.A.
in size and form. Its insurance and banking institutions vie with the C.P.R.
and Hudson's Bay Company of Quebec as giant corporations which
ultimately control Canadian economic development.

THE BASIS OF ENERGY

Given the products of the land and a labour force (which in itself may
represent a considerable market), the next basic requirement is a source
of energy. Metropolitan Canada experienced delayed industrial development partly because it had no coal. Indeed, the *Geological Yearbook of
Canada for 1863* included a special chapter on the industrial use of peat. The
St. Lawrence lands drew their initial supply of coal from the Maritimes;
the Lakes peninsula looked south to the Appalachian fields. Hydro-
electricity provided the first alternative source of energy, to be developed
and administered by the Hydro-Electric Power Commission of Ontario.
Three primary sources of water power are integrated: the succession of
schemes on and around the now fully utilized Niagara River, installations
along the Gatineau and Ottawa rivers (made accessible by improvements
in long-distance transmission) and the supplies of the St. Lawrence Seaway.
Quebec province, which claims the largest hydro-electric resources of any
province in Canada, has its principal schemes along the edge of Laurentia
—from simple units such as the Montmorency Falls in the neighbourhood
of Quebec City (Plate X), through the multiple plants along the St.
Maurice and the complex installations at the head of the Saguenay, to
immense Shield-land reservoirs with sonorous names such as Cabonga
and Baskatong (on the Gatineau). Oil and natural gas supply a third
source of energy. Both exist in small quantities in western Ontario, but
as fuel for energy they are now pipe-lined from Alberta. Montreal's oil
supplies derive almost exclusively from overseas, and the trans-Canada
pipe-line system is not extended to it. With the rise of its industries and
urban areas, metropolitan Canada has several times been upon the brink
of a power crisis. Remembering the abundance of uranium in Canada,
it is not surprising Ontario should also turn to nuclear generating stations.
There are three in the province.

The industries of metropolitan Canada may be divided into those which are based on local raw materials and those which import their raw materials from outside the area; those which spring from local initiative and those which have arisen through foreign investment or enterprise; those which look first to a domestic market and those which depend on sales abroad.

The processing of farm produce is based principally upon locally produced raw materials. It has a 'Middle Western' complexion, with meat processing, cheese and butter manufacture, milling (including corn starch and corn flour) and vegetable-oil distillation; but the scale of operations is smaller than across the border. Despite increased canning activities, the fruit and vegetable lands have not fulfilled more than modest promise. Seasonally operated factories receive lorry loads of produce and surrounding inhabitants may bathe in an atmosphere of vegetable soup and fruit stewing for several weeks on end; the uniquely flavoured peach is not commonly marketed at great distances; apples attain a perfection in appearance rarely seen in Britain. Apple juices are offered as 'cider' to the uninitiated. Tobacco, albeit of a different quality from that of Virginia, is basic to cigarette and tobacco industries which are principally located in Montreal. The product of the 'wineries' has hitherto had an unnecessarily lowly reputation, but both production and consumption are increasing rapidly. When the sap rises in March, Quebeckers contrive to extract maple syrup worth several million dollars annually; the industry is less important for Ontario. The agriculturally based industries have problems at large and in detail. The domestic market is too restricted for the national output: but sales beyond Canada have to face the considerably lower costs of other producers. During the second world war eastern Canada replaced New Zealand and Denmark as a source of supply for the British market.

Imported raw materials are the basis of most of the heavy industries of metropolitan Canada. Some refinement may take place before raw materials are shipped in from the northlands or from overseas, but for the most part industry is of primary or secondary character. This activity takes place in several different settings. Isolated plants are relatively rare in metropolitan Canada, though the radium refinery at Port Hope provides an example. Representative Canadian industrial complexes are found at the head of the Saguenay (with softwood processing and aluminium smelting in juxtaposition), along the Welland Canal in Port Colbourne, Welland, St. Catherines and Thorold (with machine shops, metallurgy, milling, softwood processing and paper manufacture). In each case a cluster of interrelated industrial settlements with a population of several tens of thousands is the outward expression of local enterprise. Much heavy industry is associated with the larger cities. Montreal and Toronto both have immensely diverse industries. Montreal's river port site is a gathering ground for primary industry: it smelts ores from the north country, refines oil and sugar from imported raw materials, has engineering and ship-building yards and is on the threshold of iron and steel production. Hamilton is unique amongst Canada's industrial cities, with its heavy dependence upon steel, and its reliance upon distant sources for coal and iron ore. Production is centred in three principal plants, but there are many lesser factories which employ steel as their raw material. Their products range from wire goods to farm machinery and passenger lifts. Hamilton, an outlier of the iron and steel complex of the north-eastern U.S.A., has most of Canada's 7,000,000 tons p.a. ingot capacity and about half of its output in value.

In addition to heavy industries and those based upon agriculture, metropolitan Canada has many consumer and transport industries. Some are the result of native enterprise: some are branches of imperial and foreign concerns. The consumer goods are largely for the domestic market, e.g. tobacco, wines and spirits in Montreal; textiles and leather goods in its surrounding plainland towns; clothing, foodstuffs and plastics in Toronto; furniture and wood-working in Stratford, London, Galt and Kitchener. Electrical goods are widely manufactured in response to an economy which revolves around its grid system; while western Ontario claims in Brantford the home of the Bell telephone company. Toronto's industrialized waterfront is heavily populated with branch plants of British companies. The American boundary also plays its role in industrial location. Thus the St. Catherines' area assembles goods, the parts for which may be imported from the U.S.A. and exports semi-finished products for completion across the border in order to escape American import duties. In the assembly of motor vehicles Windsor, Oshawa and Bronte have plants with an output of several hundred a day. The war years witnessed the rise of aircraft manufacture around Toronto and ship-building along its waterfront. Kingston (French Fort Frontenac before English Fort Henry and Fort Frederick) matches its military schools with a century-old university, and blossoms as an aluminium town.

Administration also gathers together population. Ottawa, to which the federal capital was transferred from Kingston in 1856, has its industrial complement in Hull across the river. No city in metropolitan Canada has had a more distinctive evolution than the ex-centric regional capital of Quebec. It has lived three lives: that of a fortress outpost; of a maritime and commercial centre which was the gateway to the mid-nineteenth-century colony; and of a manufacturing town dominated by the impressive skyline of the administrative and ecclesiastical buildings built upon Cape Diamond.

THE TRANSPORT NETWORK

Industrial efficiency depends to a large extent upon an effective transport network. The coming of the railway and the Great Lakes steamship first underwrote industrial opportunities. Already in the days of post coaches, which anticipated the rail routes in their axis from Windsor to Quebec City, metropolitan Canada had looked to its waterways. The 325-foot difference in water level between lakes Erie and Ontario was first challenged when the Welland Canal was born in 1829. Enlargement and reconstruction has left behind a legacy of earlier channels on the face of the land. The present waterway, which is 25 miles long and has a 30-foot draught, surmounts the escarpment in seven locks. A second routeway, linking Trenton to the Georgian Bay by way of the Kawartha Lakes and Lake Simcoe, never achieved commercial consequence. The Rideau Canal, linking Kingston to Ottawa by way of the Rideau Lakes, was a strategic routeway (in which even the Iron Duke expressed interest) that was never put to the test. It was natural that there should be concern over the physical obstacles along the St. Lawrence River. Neither capital nor technique permitted an early solution to the problem. In any case canalization introduced international considerations. A series of modest canals and locks offering a draught of 14 feet was therefore constructed on the Canadian side in the 1880's. Vessels of several thousands of tons were thus given access to the Great Lakes so that the flags of Europe, if not of the world, were fluttering in the harbours of metropolitan Canada in Victorian times.

The St. Lawrence Seaway (Figure 54), a subject of debate for two generations, was opened in June 1959, having been constructed and financed jointly by the U.S.A. and Canada. It is a multi-purpose project, with power developments aimed at meeting the shortages of Upper New York State and northern New England, as well as of metropolitan Canada. Between Ogdensburg, N.Y., and the approaches to Montreal,

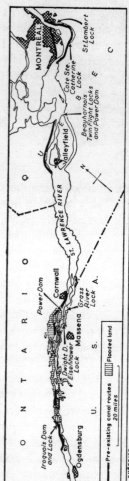

Figure 54 *The St. Lawrence Seaway*

Between Lake Ontario and Montreal, the course of the St. Lawrence River is interrupted by rapids and waterfalls. Access to the Great Lakes from the open sea has been transformed by the canalization and control of the waterway. Other control points in the Great Lakes system of communication are identified on the accompanying map (after *Atlas of Canada*, 1957, map 80) the river falls 224 feet in a distance of 115 miles. The Seaway has provided a 27-foot channel which gives the bulk of the world's shipping potential access to the Great Lakes. Prolonged American-Canadian rivalries and sectional jealousies delayed its construction. Alternative routes were proposed in former times—an American route from Oswego via Oneida and the Mohawk to the Hudson: a Canadian route from Georgian Bay via the Ottawa Valley. It was argued that existing railroads offered adequate transport, and commercial effects on different parts were anxiously debated, e.g. the possible loss of trade by Montreal and Atlantic coast ports to lake-shore rivals. A primary strategic and commercial argument sprang out of the iron and steel interests of Ohio and Illinois, which were faced with declining yields of Lake Superior ores. The necessity of cheap water transport to move Quebec and Labrador ores was a strong favouring argument. No less interested were the automobile industries of the Middle West. Certainly the Seaway need not restrict the development of Montreal, formerly the head of navigation, though it may change its course. It cannot fail to strengthen and sustain the industries of the Lakes peninsula.

This entire system of waterways ceases to be of commercial value for up to half of the year. The impact of winter on commodity movement calls for its own industrial adjustments, the substantial stock-piling of industrial raw materials during open water and of finished or semi-finished goods during the period of closure. Movement overland becomes correspondingly more important during winter, with all the costly accompaniments of road and rail maintenance. Drawing upon experience from the Laurentian valley lands, Pierre Deffontaines has written an entire volume about man and winter in a Canada where the hum of the snow plough means more than the jingle of the sleigh bell.

CONTINUITY OF LEADERSHIP

Superficially, metropolitan Canada is the most advanced, integrated and stable part of the country, yet it continues to show maladjustment both in its internal and external relations. Maladjustment is to a large extent rooted in the degree and stage of its industrialization. First, industrialization has proceeded further here than in any other part of Canada. It is claimed that 90 per cent of Canada's industrial wealth is concentrated between Quebec City and Windsor. Next, because progress tends to be cumulative, the area draws ahead faster and has a correspondingly higher power of attraction for labour. Provincial adjustments are therefore a primary problem, especially as control of investment lies largely in the metropolitan area. Thirdly, the balance between agricultural enterprise and industrial enterprise remains unresolved. Finally, the external relations of the industry of metropolitan Canada call for continuous adjustment. The farther Canada moves from primary activities, the more its products must encounter obstacles to international trading. Nor does it share those economies of scale displayed by the U.S.A., with its larger domestic market to monopoly developments. The difficulties which affect Canadian industry at large, affect those of metropolitan Canada in particular. But in comparison with most of the country, metropolitan Ontario and Quebec have a sufficiency of inherent and of acquired advantages to guarantee their continued leadership.

Canada's 'golden horseshoe', stretching from Toronto's skyscrapers to Niagara Falls' sky-high observation towers, or Montreal with its 'Expo' site flying the new red maple leaf flag, expresses a vigorous independence which strives to resist the challenge of absorption by the industrial north-east of the United States.

SELECTED READING

The detail of metropolitan Canada can be filled in from

> D. F. Putman (ed.), *Canadian Regions* (London, 1952)—Chapters 6-12

Forces behind the advance and retreat of rural settlement are investigated in two papers,

> J. W. Watson, 'Rural Depopulation in South-Western Ontario', *A.A.A.G.* (1947), 145-54
>
> J. W. Watson, 'Influence of the Frontier on Niagara Settlements', *G.R. 38*, 1 (1948), 113-19

and are complemented by

> J. Spelt, *The Urban Development of South-Central Ontario* (Assen, Netherlands, 1955)

A fine contribution to the early period comes from

> R. L. Jones, *History of Agriculture in Ontario* (Toronto, 1946)

and an important aspect of industry is reviewed in

D. Kerry, 'The Geography of the Canadian Iron and Steel Industry', *E.G.* 35 (1959), 151–63

The contribution of the seaway is dealt with by

T. Hills, *The St Lawrence Seaway* (New York, 1959)

It is natural that the best books about Quebec province should be in French. A synthesis of thirty years' acquaintance centred principally on French Canada is

Pierre Deffontaines, *L'Homme et L'Hiver au Canada* (Paris, 1957)

A slightly dated but nevertheless valuable sociological study is

E. C. Hughes, *French Canada in Transition* (Chicago, 1943)

while a valuable series of essays on current research is contained in

Mélanges géographiques offerts à Raoul Blanchard, Cahiers de Géographie de Quebec, 3,6 (Quebec, 1959)

Through an excellent system of news letters the *Association of Agriculture* keeps its finger on the pulse of adopted farms in Ste. Martine, Chateauguay County, P.Q. and Hillsburgh, Wellington County in southern Ontario

'Colonial' Canada and the Problem of the Physical Frontier

A Time of Reconnaissance—The Role of Transport—The Succession of Staples—The Want of People—A Stratagem for Development

A TIME OF RECONNAISSANCE

The northlands, shared by seven provinces exclusive of the territories, account for by far the largest part of Canada and there is good reason for treating them as a single unit from the human point of view. They are frontier country which extends from the borders of Alaska on the west to the fiords of Labrador in the east. Their poleward limits are largely *terra incognita*, politically bounded in a triangular shape and subtended by the Arctic coast of Canada. They extend through 3,000 miles as the reconnaissance aircraft might fly from the Mackenzie delta on the Beaufort Sea to Belle Isle Strait. The northlands are thus of virtually Siberian dimensions and character. Nor is twentieth-century Canada unmindful of the fact that, facing Siberia, it has acquired a political frontier in the north. This frontier is an essential element which prompts the understanding and opening up of its high-latitude lands. Canada shares many common problems with the U.S.S.R. in its Arctic domain.

The physical task of exploration alone is almost overwhelming. So much of Canada remains unknown and the northlands are known least of all. They have been photographed from the air in their entirety; but most of their territories remain untrodden. Almost anything written about them must take the form of extreme generalization. The greater part of the area is set against the background of the Laurentian Shield, with the north-western lands having a sedimentary basis, rising to the Tertiary folds of the Mackenzie Mountains. Northwards, where the land mass is broken by icy seas into an immense archipelago which cradles

the Magnetic North Pole, the fundamental geology is but little investigated. Mercator's projection may have falsified the perspectives of the
islands for a good many generations, but at least it emphasized their size.
Ellesmere Island is nearly as big as Britain; Baffin Land three times as big.
Air photography reveals not only the summer shape of the land, but also
the distribution of the main vegetational forms. The northlands embrace
three vegetational zones—the boreal forest, the birch scrub and the
tundra—before yielding to the polar ice. And the zones do not follow
simple circumpolar courses. Thus, the coniferous forest marches farther
north in the western than in the eastern half of the north country; while
there are vegetational islands reminiscent of the south in the sea of northern
forests and archipelagos of softwood in the birch scrub along river courses.

Some enthusiasts argue that in their attitude to the northlands
Canadians display a reaction akin to that of nineteenth-century Americans
as they advanced towards 'The West'. It is true for only a modest proportion. Canadians are very conscious of the east-west extent of their country;
but the north-south extent is commonly ignored. It is an aspiration of
most Canadians to journey from coast to coast; but very few have
travelled from the southern to the northern termini of the realm.
Incentives to travel north are of a specialist character, though a new route
such as the Alaska Highway can seize the imagination of the private
motorist. The means to journey from south to north are very restricted,
independent of motives. If Canada's problem a century ago was to strike
an east-west link, its task today is to integrate north and south. North-
south integration calls for a multiplicity of routes and for continuous
subsidization. If there is cause to initiate north-south routes they are
generally constructed; but there is not the same need to integrate north
and south that there was to tie together east and west.

The original north-south routes were the waterways. The watershed
of Canada's Arctic drainage is pushed deeply into the south of the
country. So much of Canada looks north or has a northern aspect. The
drainage network of most of Canada is oriented to two major depressions,
the Hudson Bay and the Mackenzie Valley. The Hudson Bay is a depression in the heart of the Archean Shield, which repeats on a larger scale the
characteristics of the Bothnian Sea in the Fennoscandian Shield. The
Mackenzie Valley (Plate XVI), excavated from younger sedimentary
series, is the continent's mirror image of the Mississippi. As Tony Onraet
reminds us in the title of his book about the Mackenzie, one goes *Down
North* by such routeways. It is no easy journey, for rapids and falls obstruct,
tributaries proliferate and named and unnamed lakes of all sizes interrupt

their courses. The Mackenzie system, changing its name at intervals along its course, is over 3,500 miles long. It incorporates the Great Slave Lake (around latitude 62°N) and drains the Great Bear Lake which lies on the Arctic Circle. Both are the size of the more familiar Lake Ontario with a water area larger than the area of Yorkshire. The Saskatchewan river system loses itself in Lake Winnipegosis on its way to the Hudson Bay; the Red River in Lake Winnipeg. Lesser rivers repeat the process. In Europe only Fenno-Karelia has such a labyrinth of waterways.

Cartographically speaking, this is a country of water without contour, for such incipient maps as have been compiled over most of the northland rarely include even form lines. Atlas maps boldly define a pattern of waterways; but the more cautious large-scale cartographer, even in relatively known tracts, frequently uses a broken rather than a continuous line for the representation of watercourses. This is not because streams suffer intermittent flow, but because their routes are literally conjectured.

The régime of the northern waterways adds to the difficulties of the mapmaker who would define them and the navigator who would negotiate them. Extensive flooding commonly succeeds the spring thaw and is exaggerated by two features. First, the thaw takes place from source to mouth: secondly, the lower courses of most rivers have remarkably slight gradients. Between Cochrane and James Bay, for example, the land falls by no more than 2 feet per mile. In the north-west, steam vessel, and barges may operate seasonally on stretches of the Mackenzie systems but for the most part the northern rivers are not easily navigable. The problem of drainage also expresses itself in the widespread muskeg, studded with lakes and pitted with water-holes. Such swamp is usually devoid of vegetation of value—a summer obstruction to overland movement and a breeding ground for mosquitoes and midges. The gently emergent clay coastlands of the Hudson Bay give rise to thousands of square miles of valueless terrain which face a muddy grey inter-tidal zone half to six miles wide.

THE ROLE OF TRANSPORT

The key to the northlands is transport. 'Roads to resources' is the accompanying slogan. Sleigh and canoe were modes of transport born of the country. The railroad has fought and (despite technological aids) still fights its way through rock and swamp, with fixed charges and high operating costs defying economic solvency. Winter ice and snow raise additional problems for wheeled transport. Construction of highway and byway is easier; but these also carry with them high maintenance costs. Roadways in the north have shown a response to strategic as well as economic forces. The Alaskan highway (Figure 55) strikes through the Albertan northland and the Yukon to Fairbanks in Alaska. During the second world war a continuous highway was constructed around the north shore of the Great Lakes between metropolitan Canada and the

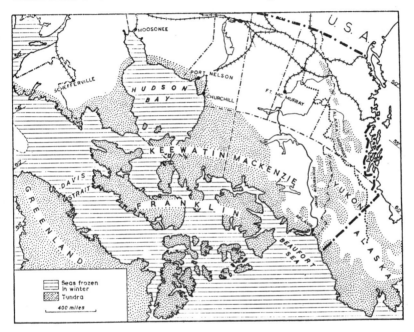

Figure 55 *Access to the Canadian Northlands*

Lines of latitude are among the most deceiving features of the map of Canada. A perspective of Canada from the north emphasizes the limited extent of the area of occupation

Red River Valley. Unless he is prepared to purchase a new set of tyres at the end of his journey, the motorist still uses the metalled highway on the American side. A Mackenzie Highway is proposed. Aircraft (using landing strips) and hydroplanes (exchanging floats for skis in winter) operate beyond railhead and road end: the fleet of privately owned aircraft is large. Most commercial organizations operating in the north country will have their own machines. The establishment of settlements such as Yellowknife, Port Radium and Schefferville has called for some of the biggest civil air-lifts in the post-war period. While private aircraft penetrate into the wilderness, international air-routes pass increasingly over it. Air-links between the cities of the Pacific and the capitals of western Europe pass 4 miles above Hudson Bay; while such a service centre as Goose Bay in Labrador is another type of specialist outlier.

Railroad, automobile, tractor, aircraft, motor launch (even canoe and sleigh) are inclined to be more complementary to each other than competitive: likewise the rare but appealing snocat and snomobile in winter. An absolute dearth of means of communication over the greater part of the northland implies that old and new methods of transport operate with a measure of unison. Most transport systems must be subsidized to a greater or lesser extent. It would not be misleading to describe the greater part of the north country as a transport desert in two senses. First, most of it is without rail communication. Secondly, the volume of local traffic moving on the railroads is negligible. Long-distance commodity movement over the northern railroads also tends to be very variable. So much of it is also one-way movement, a feature which does not commonly characterize the air-routes. Movement on the Churchill line (completed 1929) illustrates both points, while trans-Laurentian traffic is principally from west to east. Overland grain movement is naturally competitive with transport on the Great Lakes and raw-material movement from north to south is not complemented by return cargoes. Marine insurance rates, at a minimum from 23rd July to 10th October, illustrate the limited value of the Hudson Bay shipping lanes.

THE SUCCESSION OF STAPLES

Changing transport facilities have added to the variety of resources which can be economically produced (Figure 56). Indeed, a succession of staples has been called into being by the changes. Old staple products retain a place, but new come into being side by side with them. New staples also arise in new areas. Old methods of transport retain a place. Thus, the value of some commodities in relation to their bulk is still sufficiently high for sleigh and canoe to be used widely as ancillaries to the network of mechanical transport. They were the first to bear and continue to bear the rich burden of the pelteries.

Furs are commodities valuable in proportion to their bulk. Fur-bearing animals are trapped rather than hunted, trappers must be formally registered and the catch is a reward for almost scientific cunning.

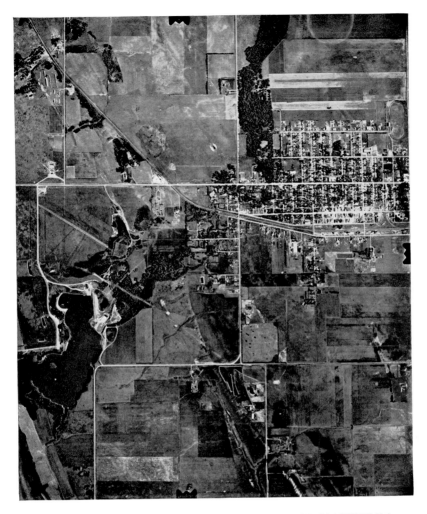

XIII THE PRAIRIE LEVELS OF WESTERN MANITOBA

Morden stands at the meeting ground of the first and second Prairie levels. This small Manitoban community conforms to the geometry of the township and range system. The area is located on the eastern part of the transect depicted in Fig. 64

XIV AN INTERIOR VALLEY OF BRITISH COLUMBIA

The interior valleys of British Columbia, isolated from each other as well as from the rest of Canada, support ribbons of settlement and cultivation on lake terrace and river terrace at the foot of steep, often treeless ranges. Penticton is a centre of one of the fruit-growing valleys

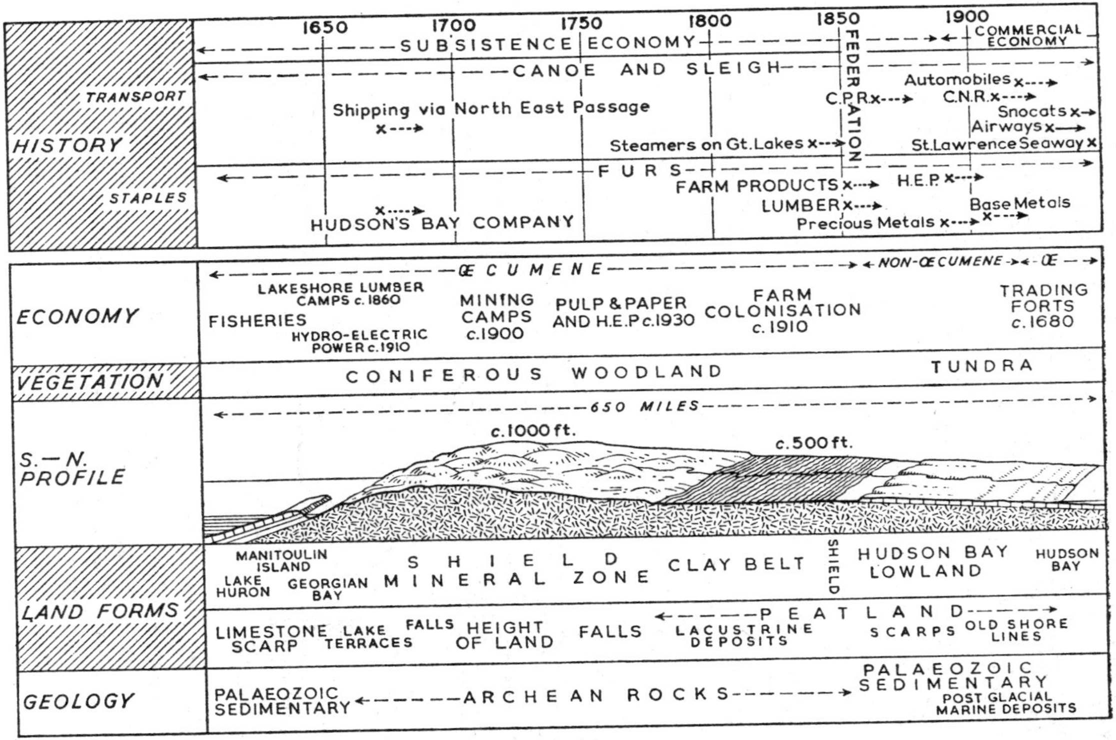

Figure 56 A Transect through Northern Ontario

Jealously guarded trapping lines, stretching through the winter wood-lands perhaps for several miles, have been established and are passed on as an inheritance from father to son. Trapping has retreated increasingly from metropolitan Canada as fur-bearing animals have withdrawn into the wilderness. Alberta and Manitoba have captured the lead from the north-eastern provinces. Returns from the forests are increasingly variable. Beaver, mink, marten, squirrel and musquash are still available; but, for better or for worse, it is the ranch-bred animal which takes precedence in Canada's fur trade today. There are still plenty of larger predatory beasts—bear, wolf, lynx, fox—the skins of which claim a price. It is even possible to go into the woodlands around such a city as Timmins (Ontario) and shoot enough black bears to provide busbies for a brigade of guards. The value of the fur harvest is maintained, but propor-tionate to other commodities it diminishes. The north country continues to offer abundant wild life for hunting expeditions. Moose and deer are stalked in the autumn and the city deep-freeze accommodates their choicer steaks.

Precious metals were the second resource to command attention. The northlands became an Eldorado, though belatedly. The gold rush to the Yukon in 1899, with Klondyke trails doubtless more Chaplinesque than Chaplin's, had their counterparts shortly afterwards with the discovery of gold in northern Ontario and Quebec. Gold-bearing ores occur both south and north of the height of land, but principally in a crescent along the southern margin of the Clay Belt. First ores excavated were sufficiently valuable to withstand winter transport by dog teams. The Hollinger mine at Timmins was the first significant development. Three years later, in 1912, Kirkland Lake was settled. In the 1920's there was an eastward extension of workings into the auriferous quartzes of Quebec at such places as Noranda, Taschereau and Abitibi, but the eastern mines have remained of more modest size than the western. The occurrence of precious metals encouraged the extension of branch railway lines northwards from the base line of the C.P.R. At the same time the Canadian Northern (later National) Railway thrust its way through the Clay Belt. This new accessibility has given a measure of stability to a fickle enterprise, although there has been considerable fluctuation in the absolute number of mines in Canada since the establishment of gold mining two generations ago. In the last 30 years the number of mines operating has varied from 32 to 148. Gold is also produced in association with other metals. So, too, is silver: which is one reason for the faltering fortunes of Cobalt. As a producer of precious and semi-precious metals Laurentia has a world status.

The mining of precious metals preceded and pointed the way to the exploitation of other ores. It was the Royal Ontario Nickel Commission of 1917 which first indicated the technical possibilities of employing a neglected resource. The extensive deposit of nickel along the Vermilion River has been basic to the rise of the International Nickel Company of Canada. Associated with the nickel is platinum. Sudbury (pop. 46,000) has emerged as the historic nickel centre and 95 per cent of Canada's output derives from its vicinity.

Quantitatively, copper is the principal non-ferrous metal produced from the north country. It is widely distributed in occurrence; but producing centres are few in number. Copper mining has pushed far north into the waste-lands, so that the mines of Sudbury and Noranda in the nearer northland are complemented (but not outstripped) by those of northern Manitoba, such as Flin Flon (pop. c. 10,000) on the cul-de-sac line from The Pas. New names constantly modify the distribution maps of known and operated non-ferrous deposits. Nickel from Thompson and Moak Lake, Lynn Lake (in Northern Manitoba) and Rankin in the North-West Territories provide examples.

The rise of iron-ore production in the northlands belongs to the last generation, for between the wars Canada had very little indigenous iron output. American supplies were too accessible and too cheap for competition to be feasible. But the threat of exhaustion of high-grade American supplies prompted prospecting in complementary parts of the Shield. The Algoma Company established a mining site on relatively low-grade ores at Michipicoten, on north shore Lake Superior. Steep Rock, in the Rainy River area west of Port Arthur, was opened up during the war years and has a prospective output of 5,000,000 tons p.a. A new scale of enterprise has been added with the exploitation of the Quebec-Labrador iron trough. A high iron content (60 per cent and more) and abundant reserves give continental stature to these deposits. The principal mining operations are administered from Schefferville, which is located some 360 miles from the Gulf port of Seven Islands (now second harbour of Canada in tonnage handled). The Iron Ore Company of Canada's Wabash Mines and the Quebec Cariter Company's mines (moving to Port Cartier) are additional developments. New railway lines have been construeted to meet the export requirements of this iron province. Major concessions have been staked out on the ore lodes extending through to Ungava Bay.

New strategic ores have pushed forward the frontiers of mining, but have found difficulty in maintaining it. Ontario's group of uranium mines in the Elliot Lake area have suffered decline; Saskatchewan's uranium is drawn from the north shore of Lake Athabasca; while Port Radium (most remote of all), which formerly refined output from the shores of the Great Bear Lake before its export 3,000 miles overland to metropolitan Canada, has virtually closed.

Softwood timber is the most widely distributed resource of the northlands. It is also a renewable resource. Softwoods have always had local value for construction and heating; but only in more recent decades have they become a third staple of wealth. Their industrial value is inseparable for transport facilities. Softwoods south of the Arctic watershed, floatable to the shore of the Great Lakes or the tributaries of the St. Lawrence, have been heavily exploited over the last 60 years. Before preventative legislation was introduced extensive tracts of more accessible timber land had been depleted, e.g. especially the north shore of Lake Superior. Softwoods north of the height of land had (and have) no more than local value until penetrated by railroad and highway. Even then, their revaluation waited on the development of hydro-electric energy. Half of Canada's softwood timber stands still have no commercial value because of inaccessibility. The new communications and power plants have enabled the rise of processing plants on the margins of and within the 'Clay Belt'. Quebec and Ontario are the leading paper and pulp producers in Canada and their raw materials derive essentially from the north country. Mills, as large as any in Canada, have been established north of the height of land, e.g. Abitibi Paper and Pulp Co. at Iroquois Falls and Spruce Falls Paper and Pulp Co. at Kapuskasing. South of the watershed, mills cluster around Port Arthur-Fort William, the northland's largest single softwood processing centre.

Milling and mining call for energy. Laurentia lacks coal and oil, though the Mackenzie Valley has abundant oil beside its one established centre at Norman Wells. But the north country is rich in water power. Quebec is Canada's richest province, having slightly more available continuous power than British Columbia. Ontario takes third place. Quebec and Ontario have the greatest installed capacity. It is over half a century since hydro-electricity first began to transform the north country, through the Kaministikwia project at Port Arthur-Fort William. In the interval the scale of production and the distance of transmission have been revolutionized. The outfall of Lake Nipigon, 250 feet above Lake Superior, has now been harnessed to the needs of the Twin Cities and supplies 100,000 h.p., while the Kaministikwia project is now elaborate with 2-mile intake tunnels. Exploitation of the Laurentian edge rivers is matched by that of the Hudson Bay tributaries. The English, Abitibi, Mississagi, Albany, Nelson and Churchill rivers all have their large-scale power plants, their projects under way or in prospect. Their energies are expressed in hundreds of thousands of horsepower. The eastern extremities of development are pushed along the north shore of the St. Lawrence to Manicouagan River near Baie Comeau, while in Quebec province power projects march steadily inland along the courses of the Saguenay, Peribonka, St. Maurice and Gatineau.

Cheap power is one of the principal amenities for rural communities, and eases the lot of the farmer. The produce of farmed land constitutes a staple output of the north, although it is not spectacular. Commercial farming has limited opportunities because of the relatively small local market, the distances from metropolitan Canada, the shorter growing season, the risk of summer frosts and the competition of more favoured areas of production. Mixed farming predominates, with grasses and grains providing a background for animal husbandry. South of the height of land, e.g. the fiorded Superior coast with its sugar-loaf mountains at the head of Thunder Bay, the environs of Fort William-Port Arthur, the western end of Lake Nipissing, North Bay, there is locally successful farming. The Clay Belt has, however, never fulfilled its agricultural promise. Pioneering still continues; but it is less commonly the homesteading type. The vigorous transformation of the primeval landscape is seen in such a settlement area as Mattagami, where scores of new parishes have accommodated thousands of new immigrants. The reverse of the coin is less easy to detect, but there is ample evidence of the retreat of the frontier of cultivation on countless other holdings.

Every resource upon which a new value is placed consolidates the movement northwards and provides a firmer basis for development. The northland is a country of hazards and risks. The yield of the pelteries is unreliable, the harvest from the land is chancy, the return of the mines must ultimately diminish. The renewable resource makes the most abiding impression—the flow of water through the turbines which is converted into light and heat: the regeneration of the well-managed forest.

THE WANT OF PEOPLE

The northlands need population to realize their resources. In general, distribution of settlement is sympathetic to transport opportunities. Thus, in the railroad lands of the 'near north' there are more than a dozen towns with more than 10,000 inhabitants and several scores with 1,000–5,000. By contrast, the North-West Territories, which frequently offer reasonable settlement opportunities, have only 20,000 inhabitants altogether and the Yukon has only 12,000. Population in the high north, which has only been enclosed with an administrative frame in recent years, is very scanty, orientated to the waterways, and grouped around the present-day counterparts of the old forts and trading stations. On the Mackenzie north of Lake Athabasca, there are encountered successively Fort Fitzgerald and Fort Smith, Fort Providence and Fort Simpson, Fort Norman and (near the Peel River confluence) Fort McPherson. The estuaries of the Hudson Bay support ports (e.g. Port Churchill, York Factory, Port Harrison) as well as forts (e.g. Fort Severn and Fort George). Beyond the Districts of Mackenzie and Keewatin, and now christened the District of Franklin, the Arctic archipelago contains only intermittently peopled research sites and mechanical weather stations.

The people of the northlands are of heterogeneous origin. Native to them are the coastal Eskimos, totalling 11,000, the best-known concentrations of which occur around Aklavik in the District of Mackenzie and Nain in Labrador. Some Indians have retreated to reservations (Figure 57); others, such as the Naskaupi and Montaignais tribes of the northeast, still hunt and trap. The first settlers were largely pioneer farmers, e.g. the mid-nineteenth-century colonists of the St. John lowlands, and the twentieth-century pioneers who spread themselves over the broad but harsh millions of acres of the Clay Belt. The Clay Belt is shared by Ontario and Quebec provinces, and there are appreciable differences between the eastern and western parts. The detail of place-names is indicative of the spread of French-Canadian settlement. In Quebec province the church has actively stimulated colonization, partly in order to offset emigration, the way to 'the graveyard for our race'. The *missionaires-colonisateurs* have carried an historic tradition of practical advice as well as of spiritual nourishment to frontier settlements, their one church absorbing immigrants of many faiths. In Ontario communities are more likely to display half a dozen churches and even more faiths. The north country contains a substantial percentage of East European immigrants. In rural areas, the heterogeneous population makes for fewer problems than in industrial settlements where differences are exaggerated through concentration. The Saddle Lake district of Alberta illustrates the ethnographic diversity of the pioneer fringe, with its small towns dominated by French-Canadian, Polish and Ukrainian elements. In contrast to the farming communities, strengthened by forestry operations, the mining and factory communities contain a restless labour supply

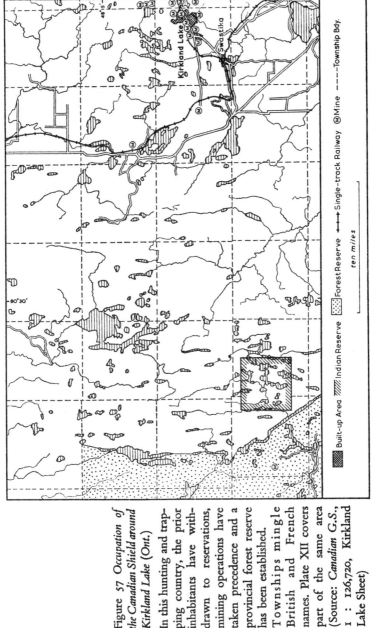

Figure 57 Occupation of the Canadian Shield around Kirkland Lake (Ont.)

In this hunting and trapping country, the prior inhabitants have withdrawn to reservations, mining operations have taken precedence and a provincial forest reserve has been established. Townships mingle British and French names. Plate XII covers part of the same area (Source: *Canadian G.S.*, 1 : 126,720, Kirkland Lake Sheet)

Built-up Area · Indian Reserve · Forest Reserve · → Single-track Railway · ⓂMine · --- Township Bdy.

ten miles

which has left scars on the social history of such places as Sudbury and Kirkland Lake (Plate XII). Labour is inherently mobile and contains much human flotsam. At the same time, the technicians and administrators are primarily Canadians by birth, rarely anything but English-speaking and often temporary immigrants from the south. Division of labour is all too often, albeit unavoidably, drawn along ethnographic lines.

A STRATAGEM FOR DEVELOPMENT

The ultimate problem of Canada's northlands is physical. There is a steady subtraction of opportunity from the 49th parallel northwards. That the land conceals riches and that the waters convey wealth is undeniable; but they never wholly make up for climatic restraints. These may be exaggerated, but frostbite, windchill and 'white-outs' are still a reality, and an instinctive awareness of these limitations checks large-scale northward migration. Secondly, there is the fact of distance, which may spell enchantment for the few but implies detachment for the many. The north country lacks strong integration within itself and is only moderately well integrated with its complementary Canadian regions. It is a land-locked country, without those facilities for American neighbourliness which Canadians equally value. As a territory of primary production, it remains a land of hard (if well paid) physical labour, though mechanization is speeding and easing the processes of extraction. Most of its developed resources are not owned by the 2 per cent of the population who occupy the roughly 3,000,000 square miles of underdeveloped land. The bulk of its not-very-carefully husbanded products are still spilled into the American and overseas markets, with some cause for jealousy of the middleman who sits in metropolitan Canada. The commercial and industrial world proclaim the new 'Empire of the North', the adventurer is wooed by the Jack London land and the artist weaned by the Tom Thomson country. Yet its new communities have conceived no regional consciousness and they seem as vulnerable to the economic controls of the south as they are to the physical controls of the north. Perhaps Canadians in their struggle with the north should adopt guerilla tactics, making their forays and assaults for short periods, extracting the most from a situation and strategically withdrawing as occasion arises. The Canadian federation, administering most of these territories as crown land, is still in process of reconnoitring them and the time is not yet ripe to produce unitary blueprints for them. It may be that the most economic stratagem for their development is to forgo the principle of permanence in occupation and to make a positive feature of impermanence for at least some of the oases of northern settlement.

SELECTED READING

There is a valuable series of reconnaissances covering various parts of the north country in

The Geographical Bulletin (Ottawa, from 1951)

Northern Canada and Alaska are dealt with in

G. H. Kimble and D. T. Good (ed.), Geography of the Northlands (New York, 1955)

P. Baird, The Polar World (London, 1963)

and

D. F. Putman (ed.), Canadian Regions (London, 1952)

while problems of accessibility from the sea are illustrated by

Ice Atlas of the Northern Hemisphere (Washington, 1946)

See also the pursuit of the theme by the octogenarian

Vilhjalmar Stefanson, The Friendly Arctic (New York, 1943)

For the north-west, see

V. W. Bladen (ed.), Canadian Population and Northern Colonization (Toronto, 1962)

C. A. Dawson (ed.), The New North-West (Toronto, 1947)

A. D. Little, Economic Survey of Northern Manitoba (New York, 1958)

F. H. Underhill, The Canadian North-West and its Potentialities (Toronto, 1959)

N. Nicholson, The Decolonisation of the North-Western Territories, in C. A. Fisher (ed.), Essays in Political Geography, (London, 1968)

The mining outpost of the north-east is analysed by

G. Humphreys, 'Schefferville, Quebec, a new pioneering town', G.R. 48, 2 (1958), 151–66

F. K. Hare, 'New Light from Labrador-Ungava', A.A.A.G., 54 (1964), 459–70

and contemporary settlement problems are treated in

G. McDermott, 'Frontiers of Settlement in the Great Clay Belt of Ontario and Quebec', A.A.A.G. 51, 3 (1961), 261–73

I. M. Robinson, New Industrial Towns on Canada's Resource Frontier, Chicago, 1962

11

The Atlantic Provinces and the Problem of 'the Poor Relation'

The Change in Status—The Environment of the Maritime Provinces—
The Tangle of the Isle—'The Land that God gave Cain'—The
Rivalry of Province and Region

THE CHANGE IN STATUS

In many ways the Atlantic coastlands of Canada suggest the poor relation. They embrace mainland New Brunswick (27,835 square miles), peninsular Nova Scotia (20,402 square miles) and Prince Edward Island (2,184 square miles) in the south; the island of Newfoundland (42,730 square miles) and mainland Newfoundland-Labrador in the north. The three former 'Maritime Provinces' are Canada's smallest provinces: Newfoundland, incorporated in 1949, is the tenth and most recent addition to the provincial fold. All share the coastlands of the Gulf of St. Lawrence as well as those of the Atlantic. In them the Appalachian and Shield country meet. They repose upon the broad continental shelf where the warmer waters of the North Atlantic drift give way to the cold offshore stream of the Labrador current. Much the same landforms which provide the background to New England are repeated along the Atlantic coast of Canada, but they are repeated northwards under progressively more adverse climatic circumstances.

There are many reasons for the relative poverty of Canada's Atlantic provinces. It is partly that they are off the beaten track. The 'Maritimes' proper are cut off by the upstanding western rim (attaining heights of 2,500 feet) which separated New Brunswick from simple access to the St. Lawrence Valley. Relative poverty is partly explained by the fact that the world in which the people live is hard—hardest of all in Labrador, 'the land that God gave Cain'. It is partly that they have not been left a rich legacy by their forefathers. Indeed ancestors have frequently misspent and misused the modest inheritance of Atlantic Canada. It is partly because the country which they occupy seems to stultify opportunity. Yet, as Ellsworth Huntington has argued so convincingly, Newfoundland is inherently richer than Iceland and Iceland is far from being a poor relation in the European fraternity. Undoubtedly, progress in Atlantic Canada is partly retarded by differential emigration, in which the more enterprising elements move to lands of seemingly better opportunity.

Yet the concept of the poor relation must not be exaggerated. For all its wealth and progress, metropolitan Canada has neighbourhoods which are as poor as almost anything encountered in the rural Maritimes. The triangle of Shield country which thrusts across the St. Lawrence to the east of Kingston provides an example. Contrastingly, the older inhabited and more mature landscapes of the Maritimes look as fruitful and well-maintained as southern Ontario. The summer idyll of Prince Edward Island, called by some 'The Garden in the Gulf' and by others 'a million-acre farm', or of the St. John river lands leading to the provincial capital and university town of Frederickton, illustrate the point. Moreover, as is evident from the backwoods of Quebec province, poor relations are not necessarily unhappy relations.

The Atlantic coastlands of Canada held a different status in former years. Perhaps reduction in circumstance as well as limited wealth affect the attitude of Canada to the Maritimes and of them to the rest of it. Their landfalls, especially the coasts of Newfoundland, were prized by West European fishermen, their outposts were highly valued by British strategists, their harbours were appreciated by those engaged in the triangular trade of sailing-ship days, their naval stores were esteemed commercially and exploited with corresponding recklessness. A hundred years ago New Brunswick, Nova Scotia and Prince Edward Island had as many inhabitants as Upper or Lower Canada. The opening of the interior brought a change in status. The rural response to changing circumstance is the theme of Andrew Clark's study of Prince Edward Island; urban response is illustrated in J. W. Watson's historical appreciation of Halifax.

The Maritimes have no precise regional parallel. New England has affinities, but its accumulated wealth and ingenuity guarantee it a continuing place in the American scene. It may have been 'bypassed' in the physical sense; but no one could describe it as a 'poor relation'. Since the end of the eighteenth century, the destinies of the Maritime Provinces have been worked out independently of New England. They live in detachment from it almost as much as they live in detachment from the English-speaking part of Canada. The paradox is that growing appreciation of the raw materials of Newfoundland-Labrador is likely to strengthen links with the American Atlantic seaboard as much as contacts with the Maritime Provinces.

THE ENVIRONMENT OF THE MARITIME PROVINCES

The inhabitants of Nova Scotia are traditionally known as 'Blue Noses'. Many have loyalist ancestors, 'true Blues' who left the U.S.A. during the Revolution and contrast in outlook with the French-Canadians who constitute as much as a third of New Brunswick's population. The nickname of the Maritimers might equally well have a physical connotation, for they inhabit a raw country climatically and one which yielded such scanty returns that eighteenth-century immigrants spoke of Nova Scotia as 'Nova Scarcity'. Humidity, the summer bane of the south-east coastlands and the winter bane of the Maritimes, exacerbates temperatures which fall to a February average of 23°F in Halifax or 16°F in Prince Edward Island. Again, the same low summer temperatures which are a boon to the increasing numbers of tourists who wish to escape the tropical heat of the interior, offer cool if not cold comfort to the predominantly rural community whose fortunes are largely dependent on them. For the Maritimes have a delayed spring, with the threat of frost breaking into a growing period, which is as short as 100 days in the interior lowlands of southern New Brunswick and 140 days in Prince Edward Island.

Romantic novelists depict the Maritimes as 'The Country of the Pointed Firs', but realists describe this southern penetration of the northern softwoods into temperate latitudes as an indicator of temperature deficiency. Thus the European ecologist arriving at the approaches to Halifax, the ria inlets of which should suggest Plymouth or Falmouth, finds himself reminded of Oslo fiord. The native of Oslo, although he would encounter the crisp crackle of winter cold, would wait vainly for the intense if brief summer heat and light in which he rejoices. There is, moreover, a climatic instability, in which summer deluge and winter blizzard may seriously disorganize daily life. Precipitation varies from rather more than 30 inches p.a. in the lowlands to over 60 inches in higher country. Snow masks the land for 5 months, and the Gulf shores from the somewhat inappropriately named Baie de Chaleurs, through the narrow Northumberland Strait to Cape Breton Island, are rimmed with ice (Plate XV). In the Gulf, patrol aircraft observe the behaviour of ice packs that drift with the tides to jam shipping lanes and bring winter movement to a close.

The habitable land of the Maritimes is much fragmented, most intensely on the eastern and western margins. The Atlantic rim attains a maximum altitude slightly in excess of 1,000 feet, and forms the backbone of Nova Scotia. Upstanding ridges of igneous or metamorphosed rocks contrast with the softer Triassic, Permian or Carboniferous basins which have been eroded between them. The Bay of Fundy is the major erosional feature, with tributary embayments around which population clusters, e.g. Annapolis Basin (Figure 58), Minas Bay. Guysborough, Antigonish and Pictou (Plate XI) are central to similar embayments on the north shore of Nova Scotia. The sandstone plain of New Brunswick is the most extensive sedimentary area. A rejuvenated drainage system, with short rivers in Nova Scotia (none exceeding 50 miles) and longer rivers in New Brunswick (the St. John River is 450 miles in length), open on a coastline in which submergence has taken precedence over emergence. Ria forms dominate, especially in the softer rock areas, but both soft and hard rock shorelines are modified by elaborate wave-built and wind-modelled features. The accessibility of the coast is important for an area which has looked into the sea to supplement the limited wealth of the land and across the sea for its markets.

In spite of physical restraints, agriculture is a leading pursuit of the Maritimes. Comparative statistics do not accord it a strong position. Somewhat less than a third of the area of the three provinces is classified as occupied farmland, but less than a tenth is listed as improved land. Prince Edward Island focused on its capital of Charlottetown is the exception. Over half of its area is improved land, the average size of holdings only a little less than in Ontario or Quebec, but the acreage of cropped land is rarely more than a third of that of their farms. Except in the most favoured areas, soil pockets are scattered, with bedrock, boulder and bogland interruptions. Arable acreage is often impossible to compact and does not lend itself readily to mechanized farming. In addition, almost any large-scale map outside the limited areas of plainland will reveal the dispersal of homesteads and the tortuous third-class roads which are the only lifelines that a lowly revenue can maintain. In the same way as the Maritimes at large lie aside from the supply area of the main continental markets, such holdings lie beyond the economic limit of produce collection. The past as well as the present dogs progress. For example, much farmland was overstocked or overcropped in former times, while woodlands

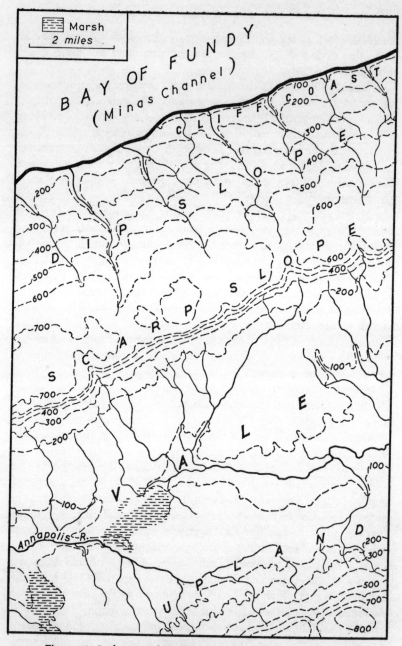

Figure 58 *Settlement and Land Forms in the Annapolis Valley (N.S.)*

Settlement patterns conform closely to the graining of the land in an area of well-distributed farms settled in the eighteenth century. Features of this

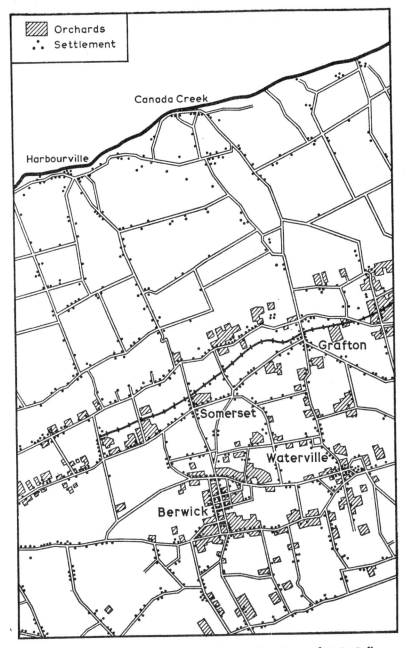

agricultural oasis still bear a close resemblance to the picture of C. C. Colby, 'An analysis of the Apple Industry of the Annapolis-Cornwallis Valley', *E. G.* I (1925), 173–97, 337–55 (Source: *Canadian G.S.,* I : 63,360, Berwick Sheet) have been denuded by the traditional, locally continuing, practice of forest grazing. Restraints on open-range grazing slowly restore a healthy forest component to the farm; and heavy use of artificial fertilizers and improved drainage programmes restore exhausted soils.

Experience has shown animal husbandry to be safer than arable cropping. Dairying predominates, despite investment costs in sturdy buildings and frequent difficulties for milk collection. Pigs complement dairy cattle: sheep play a small role. Salt-marsh grazings and historic dyked lands, such as the Tantramar marsh on the Fundy coast, supplement fodder crops. Arable land is devoted principally to fodder and root crops. In general, from French Acadians to 'Blue Noses', the farming community favours mixed husbandry. Partly in order to spread the risks that accompany specialist enterprise, they are less given to monoculture than many Canadian husbandmen. The contribution of the wood lot to the farmer's purse remains modest, although many are engaged on programmes of forest improvement. So often, the mixed farm and forest enterprise is marred by the presence of neglected timber. Around the coast, the combination of fishing and farming is practised, but the competition of professional fishermen restricts the opportunities of the casual operator with his inadequate equipment. Roots and fruits provide the basis for some rural specialization. Seed potatoes from Prince Edward Island, the Nova Scotian counties of Cumberland and Pictou, and New Brunswick's Aroostock upland hold their own as a cash and export crop. The heyday of apple exports from the Annapolis valley may be past; but 35,000 acres of orchards, which had their origin in stock imported in the 1660's, still attract pilgrims at blossom time. Strawberries, raspberries, cranberries, blueberries offer alternative, deep-frozen, export opportunities. Fur farming, established over 70 years ago, has had a more capricious career. In general, it has been a supplementary activity. The Maritimes once occupied the leading position in ranch-bred skins; but when fashion frowned on the long-haired fox, the short-haired mink replaced it only relatively slowly. The coastlands, with abundant fish for food, are well suited to mink farming. Such specialist practices help to inject vigour locally and intermittently into the rural economy. Local pioneering, principally by immigrants and restoration (e.g. through the Maritime Marshland Rehabilitation Administration), are exceptional features. In general, the frontier of agriculture shows a differential retreat (Figure 59). In few parts of Canada is a healthy agriculture more dependent upon a locally expanding economy.

What are the ingredients of this economy and how far can they con-

XV ICE BREAK-UP IN NORTHUMBERLAND STRAIT

The greater part of the St. Lawrence estuary is choked by pack ice throughout most of the winter. This photograph, dated April 22, 1960, illustrates the persistence of the ice in the Northumberland Strait looking towards Borden (Prince Edward Island) in a not abnormal season. Spring is retarded and shipping restricted by such a feature

XVI THE MACKENZIE VALLEY (N.W.T.)

The Mackenzie plain at the point where the Carcajou tributary joins the Mackenzie river from the west. The scroll-like features of the flood plain landscape resemble those shown cartographically for a portion of the Clarkesdale (Miss.) topographical sheet (Fig. 36). The approaches to the western cordillera can be seen in the distance

tinue to enrich the life of the Maritimes? The fisheries persist and by volume the Atlantic Provinces still account for a greater catch than the Pacific coast; but their contribution to the national exchequer shows a relative decline because of the enormous growth of returns from other activities. Cod and herring are the principal catch from offshore banks

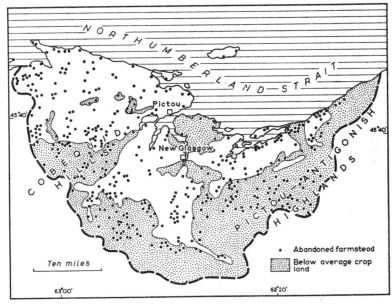

Figure 59 *Advance and Retreat of Settlement around Pictou (N.S.)*
In the Maritime Provinces there are limited tracts of rewarding farmland. Farm desertion is a feature characterizing Nova Scotia and New Brunswick as Appalachia at large (Source: N. L. Nicholson, 'Rural Settlement and Land Use in the New Glasgow Region', *G.B.* 7 (1955), 38–64, fig. 15, reproduced by permission of the author)

which mingle English and French names. The historical drying, smoking, pickling and salting operations (orientated to the West Indian market) have been complemented by canneries, deep-freezing plants, oil distilleries and fertilizer factories (Figure 60). The cultivation of crustaceans, with an eye to the American consumer, is a feature of inshore waters. Fishing still calls for some boat-building; but the historic wooden ship-building associated with such towns as Lunenburg has dwindled,

U

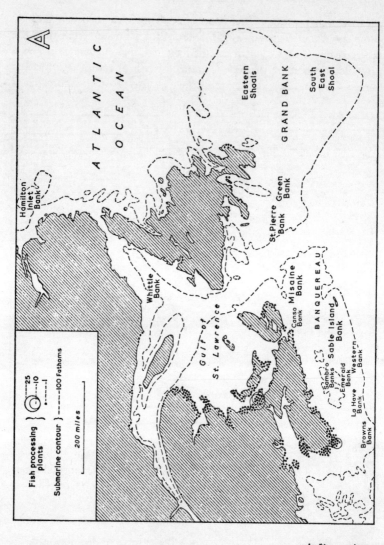

Figure 60 The Distribution of East Coast Fishing

A. Banks and Processing Plants

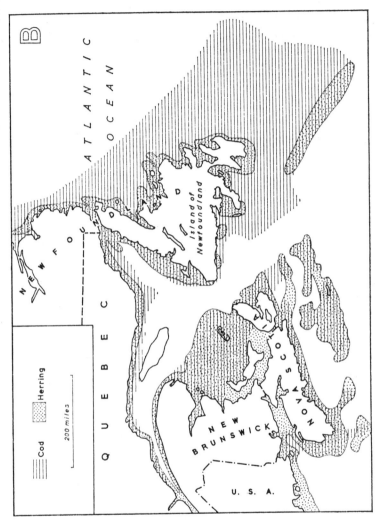

B. Cod and Herring
Fisheries
(both based on infor-
mation in *Atlas of
Canada*, 1957)

disappeared or migrated to the marine yards and naval depots of Halifax (home of the Cunard family) or St John (N.B.).

A century ago the Maritimes still drew the raw materials of ship-building from their woodlands and squandered much of their inheritance in the process. Pitch, tar, constructional pine and oak were extracted. Farmer, fisherman and merchant alike took their toll so that by the 1860's Atlantic Canada was a poor relation in the timber picture. But if constructional timbers are scarce, stands of small dimensions are fairly abundant. New Brunswick, where much of the land has been acquired for development by timber companies, has a substantial output of sawn goods and pulpwood from Edmunston, Bathurst and Dalhousie. Woodland, an actual as well as potential source of wealth, is improvable as well as renewable.

The same does not hold for minerals, and in any case the endowment is largely restricted to combustible fuel. The rise of coal mining, princi-pally in Cape Breton Island and around the Pictou lowlands, belonged to the last century. It was closely associated with the demand for fuel in Montreal and Quebec, the decline of Quebec lumber shipments and diminution in cheap return cargoes of coal from Britain, and the growth of local iron and steel production. Deposits are near tidal water, of fair quality and reasonably abundant; but competition from other fuels has impaired their prospects. Nova Scotia competes with Alberta for leader-ship in provincial output.

Primary industry is more important than manufacturing industry. Among heavy industries, the iron and steel of Sydney (N.S.) with five other related towns predominates. Ores are drawn chiefly from New-foundland. Halifax, Nova Scotia's capital and Canada's third seaport, also has iron and steel and metallurgical workshops. Apart from defence establishments and enterprises located here for national reasons (e.g. the steel and rolling-stock industries of New Glasgow-Trenton (N.S.), the C.N.R. repair shops of Moncton (N.B.), the naval yards of Halifax), two principal types of manufacturing are found—those producing high-quality goods in which transport costs are small and those which supply local wants for which distance gives protection. In general, workshops are small, rarely employing more than a few hundreds. The industrial status of the Maritimes is controlled by local markets which are too small, national markets which are too distant and international markets in which competition is too fierce. Industry in the Maritime Provinces is not keep-ing pace with that in the more progressive parts of Canada. The slower rate of development is part and parcel of the different tempo of life.

It is 70 miles across the Cabot Strait from Nova Scotia's Cape North to Cape Ray in south-west Newfoundland. The island of Newfoundland is considerably larger than Ireland: mainland Newfoundland (see p. 313) is much more extensive. The island is a structural continuation of the Maritime Provinces, oriented NE/SW, with Pre-Cambrian flanks in the Long Range on the west and Avalon peninsula in the east, and a central tract of younger rocks (Figure 61). The entire area has experienced much heavier glaciation than the Maritimes. The 6,000-mile coastline is undergoing continuous local emergence: rejuvenated drainage yields river profiles of a torrent character.

Most of the countryside is forbidding, with a background to life reminiscent of sub-arctic Norway. Much of the heavy precipitation (half of the island has 120 inches or more) falls as snow, which persists in summer on tableland or mountain areas above 2,500 feet. Land above 1,200 feet commonly displays a tundra vegetation. About half the island supports boreal forest which is but slowly regenerative: less than 1 per cent is cultivated.

Newfoundland has had a tangled history, in which there are elements of regression as well as of advance. It was a no-man's-land or any-man's land until the Treaty of Utrecht (1713). The French did not relinquish claims to coastal rights until 1904. Place-names such as Port aux Basques still recall Biscayan connections, while St. Pierre and Miquelon, off the coast of Fortune Bay, are relics of the French empire in North America. British settlement grew most swiftly in the eighteenth century: Irish increased in the nineteenth. Responsible government was given in 1854. Newfoundland had an eye on confederation in the later nineteenth century, sought re-incorporation as a Crown Colony with Britain in 1933 during the near-bankruptcy of the depression years and was eventually incorporated as the tenth province of Canada in 1949.

Newfoundland has a population of about 415,000, of whom 8,000 live in Labrador and about a fifth in the provincial capital of St. John's (pop. 77,500). Population has a high rate of reproduction and a fairly high mortality rate. It has a distinctly peripheral distribution, in response to somewhat more favoured emergent coastlands, the better woodland growth at lower altitudes and above all to accessibility. Village or hamlet settlements occur at intervals of a few miles around the coast. Among larger settlements, the specialized towns that process the island's raw materials take precedence. Corner Brook (pop. c. 23,000) in the south-west,

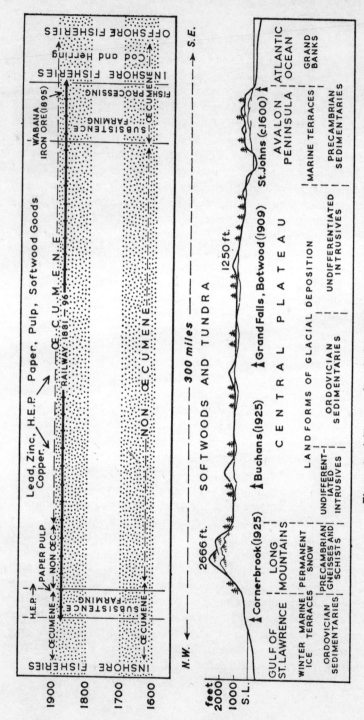

Figure 61 *A Transect through the Island of Newfoundland*

second town of the island, and Grand Falls on the interior plateau, provide examples of settlements based upon large-scale softwood milling. A skeletal railway system binds a minority of the settlements together. Bad though most roads may be, they integrate the land as automobiles multiply. Motor-driven boats have also revolutionized contacts by water.

There are, however, fundamental human difficulties. First, a pronounced stream of emigration carries away many of the younger and more energetic islanders to the mainland. This negative process of selection is one of the island's most unsatisfactory social phenomena. Secondly, the standards of health of rural Newfoundlanders are lowly. Medical reports have done much to draw responsible attention to under-nourishment in the widely scattered communities. The shortage of fresh vegetables (especially in late winter and spring), unavailability of fresh milk and monotony of diet are exaggerated by absence of sunshine. Sociological surveys of the annual round of life among coastal inhabitants indicate extended periods of unemployment and under-employment. These are in some respects a reflection of the seasonal fishing operations, but they are also a response to the malaise that accompanies malnutrition. Fishing settlements with such appealing names as Bumble Bee Bight, Little Paradise or Heart's Content leave much to be desired.

Agricultural activity is not impossible in Newfoundland and explanation of its shortcomings must be sought in lack of continued initiative as much as in physical facts. It is subordinate to forestry, fishing and mining but can be a most valuable complement to them. Potatoes and grass can be satisfactorily grown to provide the basis for animal husbandry; but to be a satisfying pursuit farming must be positively promoted. Promotion takes several forms. The central authority has initiated education and experiment (through a government experimental college), encouraged domestic industries such as weaving and furniture making, made available new seed materials and fertilizers, and aided marketing schemes. Special rehabilitation work has now been under way for more than a decade in the Avalon peninsula. Around the Bowater enterprise at Corner Brook farm plots may be run side by side with part-time factory employment. Such a reservoir of labour is important for winter felling, which may require up to 5,000 men. Grass crops stored as hay or ensilage are needed not only for local dairy herds but also for logging horses. On the company farm, beet, barley, buckwheat, turnips and orchard crops are grown. Even corn and pumpkins may ripen, though it must be admitted that the extreme south-west of the island is favoured by both a Carboniferous bedrock and inclusion within the 57°F summer isotherm.

Agricultural enterprise is probably of greater value therapeutically in mid-twentieth-century Newfoundland than as a money-making activity. Trade tables indicate its negligible role in the export picture, where the most impressive feature is the contribution of the woodlands. The historic fisheries retain their status and have been turning increasingly to new methods of harvesting. The inshore herring fishers, the offshore trawlers, the schooners bound for the banks and the summer journey to Labrador grounds remain the over-riding concern of a large proportion of Newfoundlanders. Yet paper and pulp are the leading exports and their value exceeds that of the fisheries more than twice. Mines yield a return only half that of the fisheries. The modern paper and pulp enterprise is as much concerned with the conservation and improvement of forest resources as with their commercial conversion. Its roots are already over two generations old—the Anglo-Newfoundland Development Company first establishing itself at Grand Falls in 1909. The size and ramifications of the industry give it primacy in development schemes. Softwood interests have harnessed Newfoundland's water power to their needs, with the Deer Lake scheme of Corner Brook as counterpart of the turbines on the Exploits River at Grand Falls. Newfoundland's coal resources, around St. George's Bay and north of Grand Lake, are large, but most of the fuel is in thin and fragmented seams. Some copper is derived from inland deposits around Buchan's River. The Wabana hematite deposits, on Bell Island, worked since 1895 and once employing a thousand men, no longer operate. They provided employment for about 1,000 men. Ores, of 50 per cent iron content, are exported mostly to Nova Scotia and the United Kingdom.

Newfoundland is a problem area which has yielded its harvest of government publications—from the substantial Royal Commission Report of 1933 to the 300-page document covering its incorporation as a federal province of Canada. Since 1950 the Newfoundland Industrial Development Board has aspired to plan development, but most would agree that Newfoundland's prevailing economic difficulty is a chronic shortage of capital. Since, however, most capital must derive from richer neighbours, Newfoundland is continuing aware of outside controls in its internal affairs. It has something of the character of 'John Bull's Other Island'. Retreat from independence was a means of insurance, but it implied that the tangled affairs of the island must henceforth be viewed partly from the perspectives of Ottawa. Newfoundland is in theory an asset, but in fact a responsibility. One of the basic problems is the psychology of many islanders. It may be unscientific to reflect upon the influence of environment upon the mood of men—that is perhaps because science has not probed deeply enough into this issue. When means have been created to assess the effects of environment on the energies and attitudes of men, Newfoundland will be a good laboratory for experiment.

'THE LAND THAT GOD GAVE CAIN'

Newfoundland has a mainland half, the third component of Atlantic Canada. It extends from Cape Chidley on Ungava Bay to the Strait of Belle Isle in the south, and lies essentially between the height of land and the ocean. Although a formal boundary between this strip of territory and the province of Quebec was not drawn until 1927, the use of its coast as a summer base for island fishermen has been of long standing. The mainland area is about the same size as the three Maritime Provinces with Newfoundland. The heavily denuded Labrador Mountains reach their greatest heights in the north, where a concordant range of summits exceeding 6,000 feet faces the Labrador Sea. The most impressive feature along the hard, fiorded coast, where Eskimo place-names crowd out English, is the Hamilton inlet. Hamilton River, gathering together the waters of an extensive lake system along the Quebec border, plunges over a 300-foot precipice believed to be as rich in power as Niagara and, before issuing into the sea, broadens to form Lake Melville. The Lake Melville area is the only reasonably habitable area in Newfoundland-Labrador. The 53rd parallel passes through it and also traverses Caernarvon Bay in Wales.

Climatologists have described mainland Newfoundland, with the contiguous Shield country of Quebec, as a quasi-extension of the polar region into middle latitudes. The short and cool summer does little to redeem a long and severe winter which piles up an extensive ice rampart along the coast. The high precipitation falls largely as snow, the growing period around Lake Melville averages about 90 days. As a consequence, much of the country is tundra, though more recent surveys suggest that the softwood forests are as extensive as those of Newfoundland. Fire and disease have wrought havoc in them and they are of relatively poor quality; but they represent a wholly unexploited resource.

No more than a handful of people occupy the land. It is a land that calls for specialist adaptation. Native to it are the coastal Eskimoes and the interior Indians. The rhythm of Eskimo life, with its intimately adjusted shoreline culture, springs from sealing and fishing. Nain is the administrative centre of the Eskimo country. There are two Indian tribes, the Naskaupi to the north and the Montaignais to the south. Neither has a very evolved culture: both still engage in hunting and trapping with the caribou as basic to the economy. Before the significant intrusion of outsiders over the last century there was a clearly differentiated economy. Newfoundland fishermen are intrusive upon the coast and have disturbed the Eskimo habitat. Some thousands of permanent colonists from Quebec as well as from the island attempt to eke out a fisher-farming existence. Together with the professional trappers, who extend their lines back from the St. Lawrence coast, they have intruded upon Indian as well as upon Eskimo territory. The effects have been distinctly negative. Moreover, the consequences of several generations of near-starvation, poverty of association and mental stimulus have robbed the colonists of their vitality and their settlements of development. They must be reckoned among the poorest of Canada's maritime fraternity. Mission centres and Hudson's Bay trading posts are improving elements in a rather desolate world. In Jacques Cartier's words, this is 'the land that God gave Cain'.

Newfoundland-Labrador is now a Canadian responsibility. For a variety of reasons it is brought increasingly into the *oecumene*. During the second world war, like Newfoundland, it acquired strategic importance as an aircraft staging post on the North Atlantic route. Goose Bay on Melville Lake (with some 5,000 personnel) remains a lively port of call. Air transport has given Labrador a new human accessibility; but, with only a few months of open water, and iceberg hazards (despite careful marine watch), it remains off the commercial map. Secondly, the discovery and development of large-scale ore deposits in adjacent Quebec, through the Ungava trough to Hopes Advance Bay, have indicated the mineral potential of Labrador. The railroad to Schefferville by way of west Labrador, opens a backdoor to it. A 300,000 h.p. hydro-electric plant has been established at Twin Falls on the Churchill River to provide energy for local needs and a surplus for export.

Formulae for development have been much debated. A minority view contests that its native peoples should be protected from outside interference and that the example of Danish paternalism in Greenland should be followed for the greater part of Labrador. Contrasting views suggest that there should be planning at the highest possible level with long-term and massive investment. Cheap power would certainly facilitate the refinement of minerals and softwoods, while a considerable chemical industry might be built up through nitrogen fixation (as at Rjukan in Norway) and calcium carbide (based on local calcareous rocks). The Old World—in Kiruna, Syd Varanger, Nikkeli or Imandra—provides examples of established communities which thrive in equally adverse circumstances; though Ilmari Hustich looks to Petchora for a climatic analogue. Nor, given modern husbandry, should agriculture fail to provide certain necessities of life on the spot. The Lake Melville area, with summer cultivation conditions not inferior to those in most of Newfoundland, is wholly capable of growing fodder crops to produce meat and milk. Agricultural potentiality is distinctly superior to that of West Greenland and is probably better than that of Greenland at the time of its medieval florescence.

If Canada were to employ all those modern technical innovations which represent the sophisticated counterpart of the native inhabitant's intimate adaptation, it could create a *lebensraum* out of Labrador's wilderness. In the final place, however, Canada has too many areas which in the stage of their evolution parallel Labrador. All cannot be developed and there is much to be said for a policy which sets aside some of them as refuges for those palaeo-Arctic peoples who were their first inhabitants. The alternative is a policy of piece-meal concessions which in the long run might harm Labrador more than they benefit the nation.

THE RIVALRY OF PROVINCE AND REGION

Geography poses plenty of local and regional challenges to the Atlantic Provinces and there has been no dearth of solutions. Yet the will for ultimate completion seems to be lacking. It is all too easy to single out symbols among the unfinished schemes which have been projected for the Maritime Provinces. The power of the Fundy tides has been a constant source of inspiration, embroidered upon locally in the Passamaquoddy Bay scheme on the Maine-New Brunswick border, the tidal-bore project on the St. John River and in recurrent lesser plans to harness the rush of waters on the Petitcodiac below Moncton. Canals have moved the Maritimes. Halifax harbour was linked to inner Minas Bay by the Shubenacadie Canal, but only one ship is reputed to have passed through it. Proposals to link Northumberland Strait with Chignecto Bay are periodically aired. The most fantastic conceived a 12-mile isthmian railroad, with terminal slipways, for the transport of freighters. Rusted rails and relics of a roadbed may still mark the beginning of this unfinished scheme. The winter isolation of Prince Edward Island, partly overcome by air transport and icebreaker ferries, still prompts debate upon the rival advantages of a viaduct or a tunnel to cross the 9 miles of the Northumberland Strait. Local rivalries, unequal national competition for capital, dispersal rather than concentration of energies—all probably account for battles lost in good causes.

The Maritime Provinces have not infrequently lapsed into provincialism. Awareness of a broader unity and responsibilities is expressed in the Atlantic Provinces Economic Council (1958). This practical body brings together the representatives of more than 2,000 private and corporate organizations from Newfoundland-Labrador as well as from the three original members of the confederation. The A.P.E.C., a counterpart of and a collaborator with the New England Council, inspires a new regional approach to the problems of the seaboard. So, too, does the Atlantic Development Board initiated in Ottawa. Unification is the keyword that is used among the four Atlantic Provinces in contrast to the separation which moves adjacent Quebec.

SELECTED READING

Detailed information is provided by
> D. F. Putnam (ed.), *Canadian Regions* (London, 1952)—Chapters 2–5

and there are a number of original papers about the Atlantic Provinces in the
> *Geographical Bulletin* (from 1951)

A broader framework is outlined by
> S. E. Saunders, *Economic History of the Maritime Provinces* (Ottawa, 1940)

into which may be fitted the following historical studies
> R. L. Gentilcore, 'The Agricultural Background of Settlement in Eastern Nova Scotia', *A.A.A.G.* 46 (1956), 378–404
>
> A. H. Clark, *Three Centuries and the Island* (Toronto, 1959)—Prince Edward Island
>
> J. W. Watson and A. Miller, *Essays in Memory of Alan Ogilvie* (Edinburgh, 1959)—Halifax

The most substantial work on Newfoundland, edited by a Finnish geographer, contains a thousand items of bibliography
> V. Tanner (ed.), *Newfoundland-Labrador* (Cambridge, 1947)

The Institute of Social and Economic Research of the Memorial University of Newfoundland and St. John's has published a series of monographs on the strategy for the development of the Province. Representative of them are
> Milton M. R. Freeman, *Intermediate Adaptation in Newfoundland and the Arctic* (St. John's, 1969)
>
> Cato Wadel, *Marginal adaptations and modernization in Newfoundland in Outpost Fishing Communities* (St. John's, 1969)

The Association of Agriculture maintains contact with a farm in King's County, Annapolis Valley, Nova Scotia

12

Cordilleran Canada and the Problem of Underdevelopment

A Country of Compression and Concentration—Population Deficiencies — Resources in Prospect — The Incidence of Location in Underdevelopment

A COUNTRY OF COMPRESSION AND CONCENTRATION

Cordilleran Canada, divided administratively into British Columbia (359,297 square miles) and the Yukon territory (205,346 square miles), is a country of strong longitudinal graining. It is a territory of much greater physical compression and concentration than the contiguous part of the U.S.A. The coastal valleys of Washington, Oregon and California are lacking, while the spacious plateaux of the Basin and Range Province have more modest counterparts.

From west to east the 400-mile profile through the province exhibits several distinct components. The island zone is dominated in the south by mountainous Vancouver Island, 250 miles long and as large as Ireland; in the north by the much more fragmented archipelago of the Queen Charlotte group. The islands represent the Canadian equivalent of the coast ranges of Pacific U.S.A. The Strait of Georgia and the Hecate Strait, with their related offshore banks, are the Columbian counterparts of the Willamette-Puget Sound lowlands. The coast ranges of British Columbia are a northern continuation of the Cascades of Washington State. They are a formidable mountain complex, rising to altitudes of more than 10,000 feet, though more commonly averaging 4,000–5,000 feet. From the Fraser estuary in the south to the Skeena in the north they are much fiorded, with an accompanying fringe of islands to shelter a negotiable coastal fairway. Between the coastal and interior mountain chains are dissected plateaux at altitudes of several thousand feet, such as the Caribou, which form the background to the Fraser and Skeena basins. They provide the most extensive tracts of relatively level land in Cordilleran Canada. On their eastern edge they are bounded by a complex series of ranges which are dominated by the Selkirks in the south and the less impressive Cassiar Mountains in the north. The Rocky Mountains, lying beyond the great tectonic trough—the course of which is followed by the upper reaches of the Fraser in the south and the Peace River tributaries in the north—are the easternmost constituent. The ranges on either side of the Rocky Mountain Trench reach their highest altitudes in the south, where they are traversed by few negotiable passes. In brief, this is a countryside of exaggerated slopes and of limited flatland, marked with an immaturity reflecting the Tertiary orogeny and the recency of glacial retreat.

This immaturity is evident in its drainage pattern. Rivers such as the Fraser and Skeena are interrupted by rapids and waterfalls. Lakes (perhaps 500–600 feet deep) are ponded back behind the great moraines which obstruct the over-deepened valleys, such as those of the Kootenay, Arrow, Shushwap and Kamloops. There are gorges, canyons, and cascades known and unknown. They exaggerate locally the sense of compression and they emphasize the inaccessibility of Cordilleran Canada.

No natural region of Canada has such a range of climates as the Cordillera: no Canadian province such a diversity as British Columbia. Its setting in the full westerly airstream of the Pacific and the relatively warm adjacent coastal waters combine to provide a strong maritime element. But the rugged relief intervenes to modify the maritime influences profoundly. In fact, largely owing to the character of the coast ranges, the Pacific has but little effect more than 200 miles inland. The graph of precipitation traces a profile which is almost the mirror image of that of the ranges with their intervening troughs and plateaux. Thus, British Columbia offers variation in rainfall from 200 inches on the western flanks of the coast ranges to less than 10 inches p.a. in the deep interior valleys like Kamloops. Cloudiness, humidity and fog occurrence show corresponding variety in distribution. Seasonal range of temperature also strongly reflects relief and related distance from the sea. The average temperature range increases from 40°F p.a. along the coast to over 60°F p.a. in the interior valleys. The growing period exceeds 250 days in the Puget Sound islands; but less than 50 over much of the mountainous province. Precipitation at higher altitudes and in the interior takes the form of snow during winter, but the coastal tracts rarely experience prolonged ice and snow. Winter severity increases northwards in Cordilleran

Canada, so that the Yukon, detached from the Pacific by the coast ranges of the panhandle of Alaska, has winters of sub-Arctic severity.

Plant life repeats the variety of topography and climate. In no part of Canada are more plant species concentrated. They range from the moisture-loving epiphytes around the Pacific coast to xerophytes of cactus character in the semi-arid plateaux. The margins of the south-western littoral have an almost sub-tropical opulence. Familiar west European plants and trees seem to grow to twice their usual height and size. Grasses thrive on the limited flatlands, so that cricket pitches and tennis lawns develop a sward of British character. The provincial capital of Victoria is a city of familiar English roses, hawthorns and yellow broom. The coastal climate breeds a dense, almost jungle-like fringe of vegetation back from the waterfront. It is difficult to clear and hard to keep at bay. Where deciduous rain-forest yields to the coniferous woodland at higher altitudes, branches are festooned with mosses, and pine-needles form a deep, rotting carpet where for pedological reasons—combined with light deficiency—little can grow.

Woodland prevails. In theory, four-fifths of British Columbia are forested. But much land classified as woodland supports no more than a meagre timber cover. Timber is chiefly softwood and there are softwoods of many kinds. Indeed, British Columbia shares in most of the softwood varieties found along the Pacific coast. Protected northern outliers of redwood persist in Vancouver and the silvered Sitka spruce marches southwards into coastal British Columbia. The most prominent coastal softwood is the Douglas fir, a tree of monumental dimensions, often attaining heights of 200 feet. The average height of mature oak or beech trees in England is not much more than 80 feet and some idea of its corresponding height can be gathered from the Douglas fir flagpole in Kew Gardens. Latitudinally and altitudinally, it yields to red cedar, hemlock, Sitka spruce and lodgepole pine. The lodgepole pine occurs widely in the interior, where it is especially associated with the westward-facing flanks of the ranges. The drought-resistant grasses which characterize the rangelands of Idaho and Washington extend northwards into the British Columbian plateaux, and even the sagebrush finds a place. The woodland species of Cordilleran Canada, many of them swiftly regenerative, are one of its principal resources. However, the forest mantle at large exaggerates the problem of accessibility posed by the form of the land.

The reaction of the British Columbian Indian in the face of these landward features was to turn to the sea. The shore-line culture of the salmon-eating Indians, richly recorded by observers before it declined, was oriented to the harvest of the sea and the forest edge. Abundant fish supplied them with their primary needs, while timber was elaborately worked (and eventually coloured) for boat building and house construction. British Columbia was also the home of the imaginatively carved totem pole. Agriculture, however, barely existed. To clear the woodland with the simple techniques available was impossible and fire as a method of land clearance was ruled out by the coastal climate. Moreover, to till a patch of land was to invite weed growth as abundant as seed growth. The land, indeed, conspired to deny man entry. The stimulus of climate has never been quite strong enough to overcome the incubus of relief.

POPULATION DEFICIENCIES

Difficulty of access has affected both the occupation of the land and the use of its resources. Permanent settlers first came to British Columbia from the sea and settlement is still oriented towards it. The administration is from Victoria (pop. *c.* 123,000) on Vancouver Island; the principal city of Vancouver (pop. *c.* 658,000) is the principal port and the third city of Canada. Penetration of British Columbia from the interior awaited the subjugation of the plainland Indians and the construction of the Canadian Pacific railroad. Previously only trappers and hunters moved intermittently through a territory over which the Hudson's Bay Company claimed suzerainty. The late nineteenth-century occupation of the interior followed principally the lines of the railway and territories tributary to them. The railroads, in turn, sought the routes which led to the sea. The Fraser River in the south and the Skeena River in the north are the only two major rivers which break through the coastal ranges. North of the Skeena estuary, accessibility to the sea is the more difficult because of the political boundary. The long 'panhandle' of the state of Alaska is interposed between the Pacific Ocean and the northern third of British Columbia and the Yukon.

The Cordilleran province of Canada is under-populated. There are 1,400,000 inhabitants in British Columbia and 12,000 in Yukon territory, so that the area is one of the most thinly peopled in Canada. Population, moreover, is mal-distributed. It shows excessive concentration in the extreme south-west, where three quarters of the population of the province live within 75 miles of Burrard Inlet. This concentration is integrated by ferry boats, fed from fishing smacks, protected by naval bases and enriched by ocean-going shipping. Victoria administers, but economic control rests largely in Vancouver. When the first train entered in May 1887 the commercial supremacy of the island port was challenged. New Westminster was also bypassed because of the shallows of the Fraser estuary which offered only 14-foot access at low water. The site of Vancouver is strong locally and regionally (Figure 62). It has an extensive waterfront on the deep Burrard Inlet, and accessible sources of energy. It is a point of assembly for overseas shipment of diverse products from a broad hinterland and has an excess of exports over imports. Juxtaposition of America's northernmost Pacific metropolitan area carries advantages and disadvantages. British Columbia's principal population concentration cannot divorce itself from this, suffers from emigration and experiences America's influence in the detail of daily existence and of long-term planning. Indisputably, however, much wealth is derived from it.

Elsewhere, population is concentrated in areas of specialist activity and rarely shows an even distribution of any extent. Specialized activities are tied to the products of the land or sea—the lumber towns, e.g. Powell River and Ocean Falls which are tributary to the coastal forests; the mining towns such as Trail with its non-ferrous smelter and Sullivan with its lead refinery, or Comox and Nanaimo with their coals; the fruit towns which are the annexes of the orchard country, e.g. Vernon and Penticton (Plate XIV) in the Okanagan Valley; the fishing centres, such as Prince Rupert, and the resort towns. Spectacular growth is rare; there has been local waxing and waning, in the farming as well as in mining centres. Intermittent attempts to open up virgin land for planned settlement have met with variable success, e.g. Prince George after the first world war. Over-speculation has cast an occasional shadow, in the plans for Prince Rupert, for example, the hinterland of which never fulfilled its promise. Urban settlement is even more meagre in the Yukon, where the administrative centre of Dawson City is the only example.

Ethnographically, British Columbia has a more homogeneous population than most of Canada. Its inhabitants are classified primarily as of English-speaking stock. Among immigrant elements, Scandinavians are numerous, with Norwegians as the principal sub-group. There is a small oriental minority (some trucking in the Fraser estuary lands) and a residual Indian element, still occupying reserves on valuable lands tributary to the Burrard Inlet. Population increases through immigration, but British Columbia with Ontario has the lowest birth rate of any province. Population deficiency is likely to be persistent. This is an issue which acquires fuller significance when it is remembered that Canada's Pacific coast faces a western Pacific rim of high population pressure.

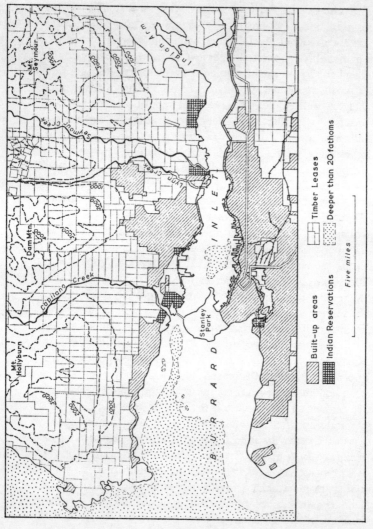

Figure 62 *Vancouver and its Environs*

The urban area of Vancouver has developed since the 1880's. Forested mountains advance to the edge of Burrard Inlet, bringing wild country to the threshold of the province's largest city. Timber and water resources are carefully conserved as amenities for the urban area (Source: *Canadian G.S.*, 1 : 63,360)

RESOURCES IN PROSPECT

Cordilleran Canada may be deficient in human resources, but it is rich in natural resources. As with much of Canada their value is frequently not fully assessed, and because of their inaccessibility many have no current value. As a result, Cordilleran Canada is very much a region of resources in prospect. It is largely an undeveloped territory and its economy is based upon primary activity. The chief sources of wealth are the forests, fisheries, mines, farmlands and water power. These have all been developed very unevenly.

The fisheries have been subject to the longest and fullest exploitation. They were fundamental to the aboriginal economy and the seashore barbecue has taken on a new form. The annual value of Canada's Pacific fisheries exceeds that of the historic North Atlantic harvest. The resource is constant neither in supply nor in regeneration. Salmon, of which there are five species, account for fully two thirds of the British Columbian fish catch by value. But the behaviour and movements of salmon are still not understood. Changes in the length of the salmon 'run', its direction or both may account for the marked changes in the distribution of fish and in the harvest during the last half-century. Returns are highly variable. Boat crews, with up to seven men operating large seine nets and expensive equipment, account for most of the catch. Canneries, formerly small-scale, seasonally operated and attracting a diversity of male and female labour, are increasingly absorbed by large organizations and concentrated in the Vancouver and Prince Rupert areas. Overproduction was a pre-war problem with American, Russian and Japanese competition affecting the markets: overfishing is a more important contemporary affliction.

Decline in yields is also a feature of halibut fishing. Halibut, drawn chiefly from the banks between Vancouver and Queen Charlotte islands, is the second catch. Herring and pilchards are basic to oil, meal and fertilizer industries: tunny is also canned. Distance from consumers has been offset by improved transport and refrigeration. Prince Rupert has a market for fish which extends to the Middle West.

Good timber, like good fish, is widely distributed and the best of the Cordilleran forests are unrivalled in Canada. But there are difficulties. Firstly, commercial exploitation is largely restricted to the coastal zone. The extended coastline opens up richly timbered tracts, but the rivers which cut back into the interior do not provide the same access as those of Laurentia. Logging operations are only differentially aided by winter snowfall; while growth conditions may give rise to an obstructive jungle of subsidiary vegetation. The size of much timber, an advantage in itself, nevertheless introduces problems of felling and removal. The delayed settlement of British Columbia and its isolation from timber markets may have restricted exploitation, but it has not suffered the exhaustion of stands common to eastern Canada. The number of points of attack on the forest and the scale of operations both differ from eastern Canada. The speed of regeneration is swifter. Sawn-wood production, pulp, newsprint, paper and semi-manufactured commodities (e.g. prefabricated houses, plywood and wallboard) have multiplied rapidly in the last generation. Large industrial enterprises are more common along the southern half of the coast. Foreign investors have moved in, so that in newsprint alone British Columbia may soon account for a third of Canada's production.

The forest has been to a large extent the enemy of the farmer. Farming in British Columbia is varied, but modest in scale. The province has about a million acres of improved farmland (not much more than Prince Edward Island in area). It is widely distributed, though extended tillage is rare. Farming in the Yukon is extremely restricted. Cordilleran Canada offers difficulties to the husbandman: land clearance the first, land maintenance the second. The density of natural vegetation and prolific weed growth of the coastlands affect arable husbandry. In the interior, the inadequacy and variability of rainfall modify farming methods, directly and indirectly. Irrigation becomes a necessity, while slope conditions coupled with the character of rainfall introduce the threat of erosion. Types of farming introduced from overseas or from other parts of Canada call for adjustments.

Commercial farming is more characteristic of the interior than of the coast. Opportunities for open-range animal husbandry in such areas as Cariboo are inferior to those in the Rocky foothills of Alberta and the Washington-Oregon grazing lands to the south. Commercial farming is nowhere more developed than in the deep, narrow valleys of the south, such as the Kootenay, Arrow, Okanagan and Shushwap. Their islands of cultivation are located chiefly upon the deltaic flats and glacial, lake or river terraces at the feet of the inner ranges. The bright verdure of lands irrigated by flooding, furrowing or sprinkling contrasts with the arid flank of the adjacent mountains and the dark pines which brush their moister higher altitudes like eyebrows. Irrigation schemes are mostly small-scale. The principal crops are apples, peaches, apricots, small fruits and vegetables. Most farms are family farms, but hired labour is expensive. The inner valleys are sensitive to isolation. The automobile has eased access to them, but they remain remote from markets. Canning, cold storage and co-operative marketing have improved their possibilities. Cold storage has extended the marketing period; the Okanagan valley alone has fifty cold-storage plants. Horticultural experiment (at the station at Sanichton, for example) illustrates the scientific approach to a pursuit which in its essentials bears comparison with that in Washington state, but which both natural and human factors restrict and restrain.

For every acre of improved land, British Columbia has at least four cultivable acres, while the experiences of high-latitude Europe suggest that the Yukon is not without potentiality. In addition to other difficulties, British Columbian agriculture suffers from deficiencies of capital. Given more generous means of water control, there could be many areas of high productivity. Given land-clearance equipment, land could be much more easily brought under the plough. The provincial land-clearance programme, initiated in 1946, established publicly owned equipment pools. Land reclamation has proceeded steadily during the last 15 years, but the end product is lost in the sea of wild land.

The other regenerative natural resource is water power. British Columbia lacks absolutely those multi-purpose projects for which neighbouring Washington is renowned. Until the Kemano and Kitimat plant focused attention on its north-west coastland, hydro-electric developments were always subordinate to those in eastern Canada. Despite its altitudes and its setting athwart the westerlies, British Columbia takes second place after the province of Quebec in Canada's power potential (Figure 63). But no province has so much undeveloped energy. Alternative sources of energy have partly restrained water-power development, e.g. coal may be used as a basis for thermal energy within sight of tumbling waterfalls. More recently, the 709-mile Trans-Mountain Pipe-line from Albertan oilfields through to Vancouver and the Pacific North-West of the U.S.A. has still further retarded interest in water-power projects. The Kemano installations were built for the specific purpose of supplying energy to Kitimat aluminium smelteries. The potentially rich Fraser River has been moderately developed. The fact that it is close to power-deficient lands south of the border has caused American industrialists to cast an envious eye on it. Capital would readily flow north to harness the Fraser and upper reaches of the Columbia system if the provincial government were prepared to relax Canadian control. Meanwhile, the only imaginative large-scale scheme looks to the remote north-east of the province where the headwaters of the Peace River are the object of speculation. Paradoxically, however, the north-eastern corner of foothill and 'apron lands' which meet those of Alberta has equally accessible oil reserves.

The presence of coal was one fact which reduced interest in water power. Resources are very variable in quality and they are scattered in occurrence. They range from semi-anthracites in thick, workable seams in the Crow's Nest area, through coking coals in Comox on north Vancouver Island, to irregularly occurring deposits in the Cretaceous series in south Vancouver Island (mined for over 100 years) around Nanaimo. They have played their role in domestic heating, in coaling during the days of steam vessels, in fuelling for steam engines and in smelting at the dispersed mining settlements. Current output is only 2,000,000 tons yearly.

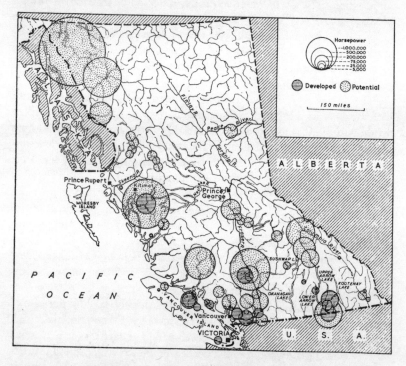

Figure 63 *The Energy Resources of British Columbia*

Most of the resources of the province are underdeveloped, not least water power. Power sites are abundant, but their magnitude calls for large-scale capital investment. Provincial authorities have been resistant to American requests to develop them (Source: *British Columbia Atlas of Resource* Vancouver, 1956, fig. 30)

Cordilleran Canada has a tradition of mining. In 1858 Canada experienced its first flurry of gold prospectors to the placer mines of the Fraser and Thompson valleys. The settlements have long since been derelict. At the turn of the century, the Yukon gave a new word for wealth to the world: Klondyke lay at the end of a trail which surmounted the high coast ranges of Alaska and penetrated into the heart of Canada's northern Cordillera. Again, it was a world in detachment from reality. Much of the evidence of the most celebrated 'gold rush' in history has been obliterated. Only a small legacy of placer-mining remains—that, and the names on the land.

In fact, Cordilleran Canada has drawn more wealth from base-metal mining than from the extraction of precious ores. Zinc, lead, copper with by-products of silver and chemicals, take precedence. The processing plants are partly located on the interior mining sites, e.g. Trail; partly on the coast, e.g. Britannia, Anyox and Belmont Inlet. The aluminium smelters at Kitimat draw raw materials from British Guiana and Jamaica for a potential 550,000 tons p.a. output.

THE INCIDENCE OF LOCATION IN UNDERDEVELOPMENT

Superficially, British Columbia is another Eden; wherever he turns, the casual visitor encounters a whole succession of demi-paradises. This larger unspoilt world persists partly because isolation has protected Cordilleran Canada; but the reverse of the coin is underdevelopment. Side by side with a landscape in which a sublime nature overshadows the works of man is to be found an economy fraught with much frustration. A scientific approach to the resources of a rich land may replace casual exploitation, a university second to none in Canada may be alert to the needs of their assessment, but the impact of 1,500,000 people upon a ruggedly mountainous land as big as France and Germany put together can never be more than superficial.

The incidence of location for this handful of inhabitants has been relaxed in three stages—by the arrival of the railhead in 1885, the opening of the Panama Canal in 1914 and the establishment of a continuous air service in the second world war. Pacific Canada now has strong mercantile relations with the seaboards of the Atlantic. It may be 2,000 miles from the centre of federal administration, but in a jet age this is only 5 hours' travel. Cordilleran Canada is the more conscious of the need for this swift link because of its setting between Washington and the state of Alaska.

It is at once aware of the penetration of American influence and of the inconsistency of this penetration. In its eyes, the strategic Alaskan highway does not follow the logical route of the Rocky Mountain trench, but was built beyond the range in Alberta. British Columbians may dismiss it as a 'tundra trail', but the dismissal probably conceals disappointment.

Cordilleran Canada may have inadequate links with the outer world, but there is more serious lack of integration within. The network of communications is thin and discontinuous. The form of the land imposes heavy costs on construction (not least on bridges) and on maintenance. Private planes as well as private cars have introduced a new measure of flexibility to movement, but the Cordillera is not good flying country. The coastlands alone have ease of access, but it is the leisured access of water communication.

Cordilleran Canada is in so many ways an outlier like the Maritime Provinces of the Atlantic seaboard: if the task of the Maritimes is to sustain, that of the Cordilleran province should be to promote. So much of British Columbia is eminently habitable. Pacific Canada is not the same as Pacific America, but there are common features. It cannot play California to the lands beyond the range and Canada's Pacific interests cannot be commensurate with those of the U.S.A.

SELECTED READING

Many local details are provided once again by

  D. F. Putnam (ed.), *Canadian Regions* (London, 1952)—Chapters 19-21

which may be supplemented by relevant sections from

  O. W. Freeman and H. W. Martin (ed.), *The Pacific North-West* (New York, 1954)

and seen in new perspectives through

  G. Taylor, 'A Study of Topographic Control', *G.R.* 32 (1942), 373-402

A visual summary is provided by

  The British Columbia Atlas of Resources (Vancouver, 1956)

which may be complemented through the successive

  Transactions of the British Columbia Natural Resources Conference (Victoria, 1956 and annually)

13

The Prairie Provinces and the Problem of Agricultural Adjustment

The Steppes of Canada—The Battle of the Grains—The Process of Adjustment — The Revaluation of Resource — A Difference of Dimension

THE STEPPES OF CANADA

'The Prairie Provinces' is one of the most popular concepts associated with Canada. Yet the phrase is something of a misnomer, for grassland is associated with only a third of their surface area. The Prairie Provinces, elongated north-south and approximately equal in area, embrace extensive stretches of the Canadian Shield and include almost as much coniferous woodland beyond their parkland fringes as they include grassland. The provinces are identified essentially with their southern parts which are in fact the longest occupied and most fully settled (Figure 64). This territory, stretching from the Red River Valley in the south-east to the Rocky Mountain foothills in the west, is curiously landlocked. The continental setting impresses itself climatically upon those who live there and spacially upon those who travel there. The long rivers, linking together lakes half the size of old-world nations, thread their way northwards or north-eastwards across an uninhabited land to a barren sea. The height of land lies almost everywhere south of the border. The artificial boundary, 800 miles of modestly defined barbed-wire fences, separates the identical landscapes of Minnesota, North Dakota and Montana.

South and west of the Shield extends a sedimentary scarp country inscribed on a grand scale. It is convenient, if an over-simplification, to define it as consisting of three levels. These broad morphological features were first identified by John Palliser a hundred years ago. The lowest step

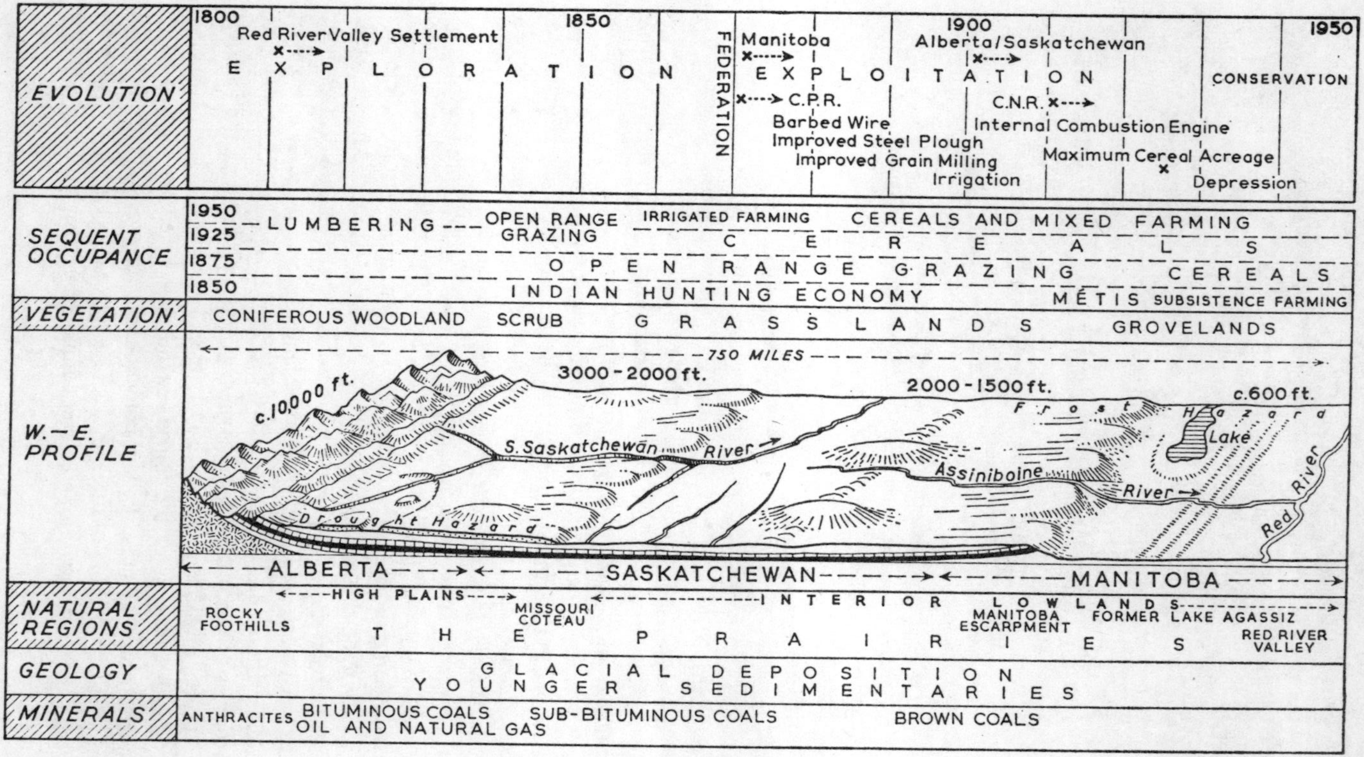

Figure 64 A Transect through the Prairie Provinces

is in the east at altitudes of 800–900 feet, and based upon bedrock of Ordovician, Silurian and Jurassic origin. Central to it is the Red River, draining into Lake Winnipeg. Most of the area was covered in the post-glacial period by the waters of Lake Agassiz. The former lake bed is extremely flat, locally swampy as in the Big Gray Marsh, and at the time of the spring thaw or after heavy rains, readily floodable. Settlements are sensitive to minor variations in relief, standing on former shore-lines and deltas, sometimes among the mixed deciduous and coniferous woodlands which march farthest south in this part of the Prairies.

The western edge of the lowest step is commanded by the Manitoba escarpment, the wooded heights of which approach 1,500 feet in the Pembina Mountain and exceed 2,700 feet in the Duck Mountain. The principal gateway to the second prairie step is opened by the Assiniboine River, with its tributary the Qu'Appelle. To the north-west is the Saskatchewan River. The broad rolling landscape, at altitudes averaging 1,500 feet, has extensive spreads of glacial clays, undrained 'sloughs', and sands (e.g. in Moose Mountain and Beaver Hills). Rivers are deeply en-trenched and broaden into long lakes where they encounter bedrock obstructions. A greater variety of relief is found than is generally believed; but limitless, almost oceanic horizons usually take control. A sea of grass prevailed at the start of its occupation. The North American steppeland, looking like the Wiltshire Downs writ large, was no welcoming land-scape. It has terrified many with its openness and claimed them as victims of prairie madness. As Manitoba is divided by the Manitoba escarpment, so the Province of Saskatchewan is divided by the Missouri Coteau which strikes across the border in a SE/NW direction. The names given to the component hills tell something of the country—Dirt Hills, Cactus Hills, Vermilion Hills; farther north, Bear, Eagle and Thickwood Hills. The third prairie step lies beyond at altitudes of 2,000 feet. This High Plain country yields in turn to the more turbulent Rocky Mountain foothills. It is traversed by rivers which rise in the Cordillera and which are tribu-tary via Athabasca and the Peace river in the north, the Mackenzie and the Saskatchewan rivers in the south to Lake Winnipeg.

A climate of extremes and uncertainties contributes variety to the monotony of the steppeland. Seasonal extremes, the impact of which is cyclically exaggerated by short-period climatic changes, are accompanied by remarkable variations in weather conditions. Averages mean very little. Two primary hazards leave their mark upon land-use distributions —frost in the north-east and drought in the south-west. The growing season ranges from a full 150 days in the south-west to not much more than

'90 on the north-eastern edge of the parkland. A rigorous winter, with the thermometer averaging –10°F in Saskatoon in January but intermittently tumbling to –30°F or even –40°F, may freeze the ground to depths of a yard or more and eliminate most outdoor activity. Precipitation is small, varying from rather more than 20 inches in the Red River Valley to less than 10 inches in the south-western parts of Saskatchewan and southern Alberta. This semi-arid triangle is still identified by Palliser's name. Most rain falls in summer, the bulk during the growing season. Relative humidity is low: the snow cover is scanty. Open landscapes give uninterrupted play to the winds, with summer dust-storms complementing winter blizzards. The warm *chinook* wind intrusive from the west plays pranks with winter temperatures, hastens the arrival of spring, exaggerates the rate of evaporation and accounts for some of the tall tales which emanate from this hard country. Yet in the climatic perspectives of the continent, the Prairies are more favoured. Lorin Blodget (cf. p 117) indicated a century ago that spring arrived almost simultaneously from St. Paul to the upper Mackenzie Valley, while a close similarity of flora from 45°N to almost 60°N intimated summer conditions not far removed from those of the upper Middle West. Summer is hot and sunny, with July temperatures averaging 65°F in Saskatoon. Humidity is low. Under such circumstances, the land can be reduced to tinder-box dryness and the prairie fire was one of the summer risks to which early settlers were subjected. Perhaps the black grassland soils found in parts of Manitoba owe their origin to fire.

THE BATTLE OF THE GRAINS

The principal problems of the Prairies have been and remain rooted in agriculture. The task of the prairie farmer is one of perpetual adjustment to the local, national and international scene. He struggles constantly with marginal conditions of cultivation, either physically or economically. Not much more than a century ago, a contributor to the *Edinburgh Review* (1857) was ascribing the failure of the Red River Valley settlement to 'defects of climate, position and communication'. The farmer could only exist if he had links beyond the grasslands and could only prosper if he replaced natural grass with cultivated grains: he can only maintain this status now if he diversifies his husbandry. For, despite the opening up of other resources, farm land persists as the chief source of wealth of the three Prairie Provinces.

The physical geography of the Prairies favours grain farming, but possibility was only transformed into reality through steam transport. Until the establishment of swift, mechanical communications, the grazier took precedence over the arable farmer. Cattle ranging was the first phase in economic evolution. From 1876, short-term leases were allocated to ranchers, store cattle replaced buffalo on leaseholds thousands of acres in extent and overstocking was a common error. The railroad, bringing in more settlers and carrying out their produce, slowly forced the cattle realm to the frontier of the Rocky foothills. In general, railroads followed settlers rather than settlers railroads, while the speed of immigration was such that settlers outpaced surveyors. Over the face of this flat land two primary patterns were laid out simultaneously—the township and range-lines which spelt out the road network, and the twin railroad system. Rail and road evolved independently, but complementarily.

Figure 65 illustrates the pattern of the railroads in east-central Saskatchewan. Two competitive lines, the C.P.R. and the C.N.R., constructed branches at approximately 20 miles distance from each other. The railroad system was originally conceived as complementing horse-drawn transport, with the '10-mile haul' as the maximum distance for grain movement. Stations or railroad halts were established almost equidistantly along the tracks. The map indicates the infrequency of double tracks—only the main east-west C.P.R. carrying a sufficiently heavy traffic to justify doubling. The flow of traffic along these lines is highly seasonal. The combustion engine has modified the role of the railroad locally, but the long-distance movement of farm produce has changed little. There remain some elements of truth in W. A. Mackintosh's observation that 'the railway . . . with its two to five grain elevators, the post office, general store, machinery shed and branch bank, closes the circuit through which the power of the world's economic organisation flows'. Visually, the line of steel, followed by successive elevators stepped out as if by a draughtsman's dividers, is a primary component of the human landscape. Rarely has an agricultural economy favoured the growth of such a distinctive transport network, or such a transport system remained so indispensable to the economy.

Other technical factors aided the revolution in the use of the prairie lands—the simultaneous fall in shipping rates, the spread of durable steel ploughs, the introduction of steel rolling mills which could grind the hard, glutinous wheat, improvement in methods of storage, mechanization of field operations. They were basic to the golden age of the prairie farmer when specialized, one-crop farming predominated and wealth was drawn

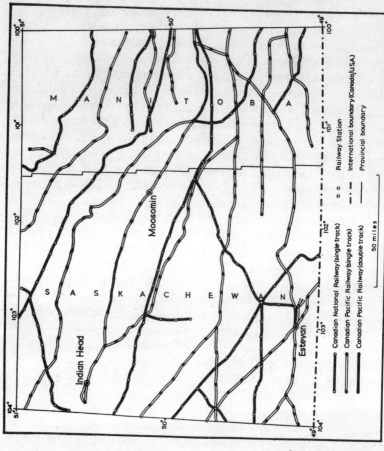

Figure 65 *Railroads in the Central Prairies*

In few parts of the world has there been devised a transport network which reflects simultaneously the economy of the country and the interplay of competition between two different railroad companies. The network was constructed before the significant rise of road transport (Source: *Canada Dept. of Mines and Resources*, 1 : 506,880)

from large-bulk, small-value commodities. But overstocking during the cattle phase was followed by overcropping during the phase of bonanza wheat farming. Uniformity of physical background might favour uniformity of crop and the store of mineral wealth in the soil might enable the cultivation of a specialist crop for a long period; but prairie soils suffered exhaustion with wheat 'mining'.

Wheat was the pioneer crop and continues to be the chief cash crop; though barley and oats are widely cultivated. Agricultural economists and geographers have identified the boundaries of the spring wheat belt (continuous south of the border) and the belts of barley and oats into which it merges. In few parts of the world can there be such a specialized interest in grains. Nor can there be many crops so dynamic in their distribution. The total area devoted to wheat and other grains has changed substantially in living memory—reaching a peak in the middle twenties and declining sharply in the interval. Within the framework of the area devoted to grain cultivation, new strains and species are continually substituted for old, specialization and diversification constantly weighed in the balance. It is a long time since Lapland barley seed and Galician grains were first imported. Subsequently, the Prairie Provinces have established seed-breeding stations which have produced special varieties with names renowned the world over, from Marquis (in 1892) and Thatcher to Regent and Redman. Battles against rust, foot-rot and lodging have been fought side by side with those against drought and frost resistance. Yields are four times what they were a generation ago and can still be substantially increased.

As a result of the rigorous climate, wheat is chiefly spring-sown, although there are limited areas of fall-sown grain in south-west Alberta and southern Manitoba. Sowing usually takes place between the end of April and mid-May; harvesting, from mid-August onwards. Because of the short season of field operations, duplication of farm machinery is unavoidable and there is heavy capital outlay. Pressure upon harvesting operations is the more acute because of the late-autumn closing of the Great Lakes waterways, the main route followed by exports. Despite mechanization, there is invariably a shortage of labour during the harvesting period, a shortage contrasting with under-employment at other seasons. The frenzy of the grain harvest takes place upon farms of different average size—200–300 acres in the Red River Valley, 300 acres and upwards in central Saskatchewan, 400 acres or more in Alberta.

Prairie farming has experienced a number of adjustments in recent years. They are a response to both physical and economic hazards, and may be observed equally in the detail of field operations and at the highest level of provincial planning. Physically speaking, the maintenance of soil quality and the improved distribution of water resources are primary.

Soil quality is best maintained by rotational practices, perhaps through the association of animal and arable husbandry. The Prairie farmer does not commonly keep stock—save for Albertan ranchers, some of whom fatten their beef cattle on irrigated alfalfa and fodder beet in the Bow and Old Man valleys. The restoration of cattle to the prairie scene, linked with new grass and forage crops, presages an accumulative phase in the cycle of soil fertility. A very restricted range of crops has hitherto been available for the prairie farmer (Figure 66). Drought, evaporation conditions or the short growing season discourage most roots and the most common crop of the U.S.A., maize. The import of Siberian clover seed has provided a much-needed legume; while pulses, flax seed and sunflower seed are also grown. The breeding of grasses such as the crested wheat grass, which can be grazed well into the autumn and is resistant to frost, has provided a new pasture base.

Pasture developments owe much to public as well as to private initiative. No single explanation accounts for the widespread farm abandonment suffered by the Prairies during the inter-war years. It was accompanied by pioneering on the forest fringes (Figure 67) and frequently provided examples of the 'hollow' frontier—the firm line of advance into the wilderness being backed by tracts of deserted farmland. Both the semi-arid cycle which gave rise to 'Dust Bowl' conditions farther south and the years of the great depression contributed to desertion. Abandoned land has been the lively concern of those responsible for administering the Prairie Farm Rehabilitation Act (1935). Land classification and the sowing of demonstration grass lots on private holdings has accompanied the establishment of 3,000,000 acres of community pastures with leased grazing on abandoned land. Parallel with this reclamation, there have been clearance schemes sponsored by the Veterans Land Act (1942) in such areas as Pasquia (near the Pas) and Brick River (near Porcupine Mountain).

Conservation and control of water resources are a second great issue. They become increasingly acute to the south-west, where frost risks are lowest. Tillage practices display a variety of adjustments. For example,

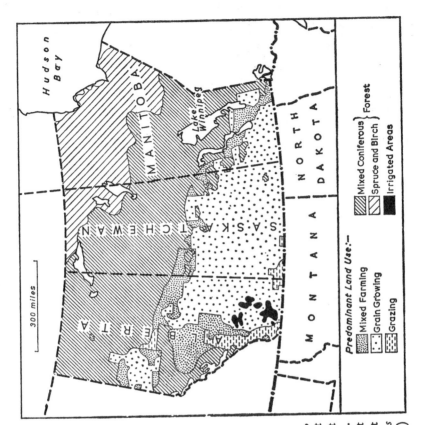

Figure 66 *Agricultural Distributions in the Prairie Provinces*

Cereal production still dominates the Prairies, with the drought hazard of the south-west and the forest hazard of the north-east restricting the classical crescent of grainland. Crop combinations in the historic wheat counties are changing as in the Middle West (after *Dept. of Mines and Technical Surveys* map of Natural Resources, 1 : 6,336,000, 1950) shallow-ploughing or disc-harrowing has usually replaced deep-ploughing, to prevent loss of surface soil through wind erosion. 'Dry farming' persists with triennial or biennial cultivation and long summer fallows, but it calls for a large holding. Strip farming, with entrenched boundaries to check soil drift, inter-tilling and shelter-belt planting all reduce soil and water loss.

Prevention of water loss is complemented by provision of water supply. The first irrigation in the west is ascribed to 1879; the first measure from which water management stems is the North-West Irrigation Act of 1894;

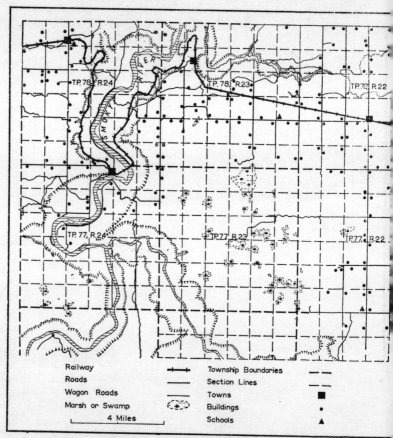

Figure 67 *Settlement in the Peace River Valley*

The township and range grid forms the basis for incipient occupation of the Peace River area. Occupation dates principally from the 1920's

among initiators of large schemes was the C.P.R. and the most extensive area first provided with water was along the Bow River. The rain-shadow country draws perennial water from the high-level Rocky Mountain snowfields. The volume of water reaches a maximum during the growing season, though while traversing the dry plains the mountain-fed streams diminish in volume. Entrenchment of rivers aids damming, but it implies that pumping must supplement gravity flow. South Alberta's Western and Eastern, and Lethbridge Northern, Irrigated Districts support ten times as many inhabitants per square mile as dry-farmed Alberta, and half an acre

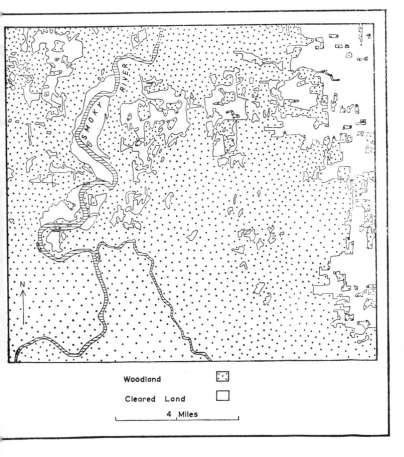

Woodland

Cleared Land

4 Miles

and earlier details are found in H. M. Leppard, 'Settlement of the Peace River Country', *G.R.* 25 (1935), 67–78 (Source: *Canadian G.S.G.S.*, 1 : 126,720, Watino Sheet)

of irrigated fodder land has equivalent grazing capacity to several hundred acres of open range. Saskatchewan's irrigated land is more scattered and its projects generally smaller than those of Alberta. Of Canada's 750,000 acres of irrigated land, 90 per cent is in southern Alberta and Saskatchewan.

The crop-risk area does not need irrigation all the time, but its yields are improved when water is available some of the time. It is believed that a full third of the cultivated area is suitable in soil, if not always in contour, for the employment of irrigation; but surveys offer discouraging estimates of water availability. Again, the scale of operations and limitation of capital have hindered development.

The Saskatchewan River has caught the eye of engineers from early times. In 1859 Henry Youle Hind in his *Report on the Assiniboine and Saskatchewan Expedition* conceived 'a dam 85 ft. high and 600–800 yards . . . across the deep narrow valley in which the South Branch flows', although his intial project was to create a navigable water link with the Assiniboine. In 1952 a Royal Commission reported on a south Saskatchewan river project and work on a vast rolled-earth dam is under way to back up the river, create a reservoir to meet the needs of 500,000 acres, establish hydro-electric power and recreational amenities. Evidently the cost of controlling the Saskatchewan River cannot be borne by the irrigated lands, and both federal and provincial governments have made their contributions. Government assistance is also forthcoming for small dams, dug-outs and local schemes of which there are tens of thousands.

In adjusting itself to risks, prairie farming has sought new outlets for its staple crops, e.g. egg and poultry production, as a means of transforming grain. As in the U.S.A., it also seeks to buttress itself through amalgamation of farmsteads. The International Wheat Conference, stabilizing the world market and allocating export quotas to participating countries, gives pride of place to Canada. The Wheat Utilization Committee of Washington looks to the expansion of markets in underdeveloped countries. There are no domestic price guarantees, but crop-failure awards protect areas of physical hazard independently of private insurance schemes, while the monopoly marketing agent of the Canadian Wheat Board (1943) is alert to all signs of economic change.

THE REVALUATION OF RESOURCE

All schemes for capital investment in the Prairies must be seen against the changing structure and distribution of population and of economic enterprise. Population is heterogeneous in origin, uneven in distribution, mobile rather than stable in location. Resources other than agricultural have undergone profound revaluation.

The population of Alberta (1,123,000), Manitoba (850,000) and Saskatchewan (880,000) is concentrated in the steppeland halves of the provinces and continues the Canadian characteristics of alignment along the American border. Population is of highly varied origin and continues to show certain well-differentiated features. The first white inhabitants of the Assiniboine and Blackfoot tribelands were in most instances English-speaking (including immigrant Americans) and in general they occupied the most fertile and most favourably situated lands. French-Canadian colonial settlements were only moderately successful. St. Boniface, a twin town beside Winnipeg, provides an example. A scatter of place-names through the clearings of the Peace River area recall the final French-Canadian phase of colonization. Other non-Anglo-Saxon groups were marginal either to the best land or to the lines of communication. East Europeans, such as Ukrainians and Poles, frequently formed the most successful colonists. In an 1899 report to the Ministry of the Interior, they were described as 'people with less craving for the society of cities and with fewer and less complicated wants . . . with skill in dealing with primitive conditions'. It might have been added that they were accustomed to the rigours of a continental climate. Homogeneous confessional groups persist, e.g. the Mennonites, of whom there are some 70,000 north of Saskatoon and others who occupy 25 townships in the Red River Valley, or the Doukhobors who continue to express their extreme non-conformity in parts of Alberta. Group settlements, reflected strikingly in the ethnographic maps of the first *Atlas of Canada* (Figure 68), have been increasingly assimilated, though at a slower speed than in the U.S.A. Legacies of original land-holding systems can sometimes be traced on the face of the land.

Population growth in the Prairie Provinces, which reached a climax in the first decade of the twentieth century, fell away swiftly after the first world war save in limited areas such as the Peace River. Between 1931 and 1951 there was a sharp diminution of population in Saskatchewan. In general three trends may be observed—a movement from country to town within the three provinces, a specific migration to the eastern and western margins of the Prairies (especially to the Edmonton and Winnipeg areas) and a general migration from the heartland to other parts of Canada.

Winnipeg (pop. *c.* 256,000), Manitoba's capital and university town, is also the oldest and largest Prairie city. It is the focus of routeways leading

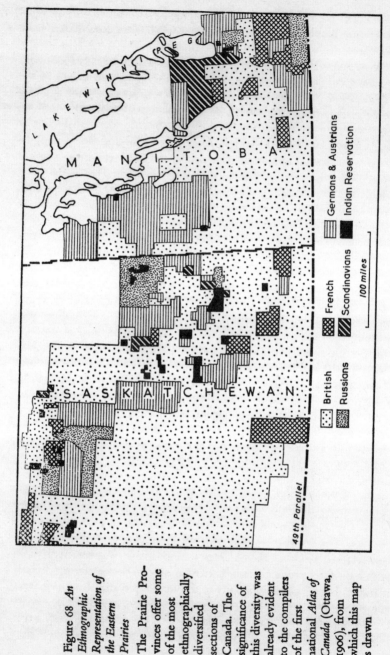

Figure 68 *An Ethnographic Representation of the Eastern Prairies*

The Prairie Provinces offer some of the most ethnographically diversified sections of Canada. The significance of this diversity was already evident to the compilers of the first national *Atlas of Canada* (Ottawa, 1906), from which this map is drawn

west by way of the Assiniboine River and south by way of the Red River. It is near enough to the hydro-electric power of Winnipeg River, the softwoods and mineral resources of the Shield area to share in them as well as to capitalize on Canada's richest grain-producing area and a province in which a third of the manufacturing plants have been created in the last decade. Although lacking a lakeshore setting, the crowded railway yards, slaughteries, meat processing and the frenzy of the grain pit suggest that Canada's fifth city is a little Chicago. The cities of the central steppeland—Regina (pop. *c.* 89,000), Saskatoon (pop. *c.* 71,000) the provincial capital, Brandon (pop. *c.* 25,000)—are principally farm-supply centres with mechanical workshops and assembly plants, specialized services from well-diggers to mail-order store-keepers, and limited processors of raw material. Calgary (pop. *c.* 196,000), on the Bow River at the gateway to the foothills and on the main route of the C.P.R. via Kicking Horse Pass, was the first urban focus of Alberta. In its environs dry-farmed grassland matches widespread irrigated land, while the annual 'stampede' reflects its debt to the cattle economy. Edmonton, eighth city of Canada, is the most rapidly expanding city in Alberta. This old Hudson's Bay Company fort, linked by the C.N.R. through the Yellow Head Pass to the Pacific in 1905, was an outpost on the way to the Mackenzie and its tributaries. The northern component in its life remains. It is the point of departure of the Alcan highway and of new railroads and new highways to the Mackenzie. Administrative and university functions, grain milling, meat packing, softwood processing, gain in importance as Edmonton becomes the regional focus of oil and natural-gas refining.

The opening up of this enterprise portends a better balance in the prairie economy, but it calls for transfers in labour and capital from an overproductive granary to undercapitalized thermal resource. For three generations coal has been actively mined in the three provinces while such towns as Medicine Hat and Viking have long exploited natural gas for local purposes. Most of Canada's coal deposits occur in the Prairie Provinces; the lignite and sub-bituminous workings of Manitoba and Saskatchewan (e.g. Estevan and Weyburn) being succeeded westwards by Alberta's bituminous fuels (Lethbridge), the excellent coking coals (Kootenay) and anthracites (Banff) of the Rockies. Alberta normally leads among Canada's coal-producing provinces, accounting for more than a third of the country's output. Its problems have been those of physical contortion in the highest-quality seams, the small scale of individual mining operations, the multiplicity of operating units, the limited local market, unskilled-labour supply, abundance of more attractive rival employment during the summer and competition from other sources of energy.

The fortunes of prairie coal have suffered absolutely from the development of petroleum and natural gas. Coal output (c. 11,200,000 tons for Canada in 1960) has halved in a decade and is subsidized. Petroleum and natural-gas production have a significant national status today. The value of oil output is five times that of coal production and meets about two thirds of Canada's domestic needs. Moreover, the prospects of additional discoveries of new fields are good. At the end of the war there were about 400 operating wells, chiefly in the established Turner Valley and Lloydminster fields. Since 1947, with the discovery of the Leduc fields, their number has increased to over 14,000, while the reserves of the Athabasca tar sands form an immense potential. Part of the petroleum is refined in Alberta; part is exported as crude oil to the principal refineries of Ontario. Refineries have given birth to new towns as well as revivifying old. Jumping Jack, near Calgary, is an example of a new settlement engaged in petro-chemical industries: Medicine Hat, of a settlement strengthened by artificial rubber production. Natural gas is also partly processed on the spot and partly exported. Pincher Creek, in the foothills, is a new town producing sulphur, propane and butane. Saskatchewan and Manitoba also have petroleum and natural-gas reserves, though the degree of development diminishes eastwards. A network of pipe-lines 25,000 miles long stretches east and west. The main trunk for crude oil belonging to the Interprovincial Pipe Line Co. extends from Edmonton via Sarnia to Toronto, and a Trans-Mountain pipe-line links Edmonton to Vancouver and the Puget Sound cities. Gas pipe-lines stretch west to Vancouver and east to Montreal. Potash deposits also claim the attention of leasing companies, with Esterhazy (Sask.) claiming the world's largest mine. These new resources give to the Prairie Provinces a valuable complement to their agriculture; while in the speed of their development and the scale of their promise, they give something of a Texan character to the economy.

A DIFFERENCE OF DIMENSION

The visitor to the traditional south of the Prairie Provinces is aware of a difference in dimension from other parts of Canada. In addition, the 'two-dimensional' character of much of the physical landscape often lacks a fourth dimension. Youthfulness characterizes both the physical and human scene. A transiency in the cultural landscape contrasts with the air of maturity found in much of eastern Canada. It is reflected in the unpainted clapboard houses and barns, the corrugated iron and the barbed wire, the tufted windbreaks which struggle to survive around the homestead, the looping telephone wires and electricity cables, the dirt-and-gravel roads which run into towns that are less compact and tidy than in the east. Some rural settlements have the appearance of Hollywood stage sets. For much of the year the farmland looks unkempt and neglected, before it breaks free and is transformed into the equivalent of a hundred Canaans. Good harvests are photographed, but lenses are shut upon bad. The landscape is workaday, there has not been time to dwell upon aesthetics. There is no place for the loving labour of the *petit propriétaire*: a giant area has called for extensive methods. Grain farming, especially, calls for bigger and bigger farms.

Human forces have not been especially well regimented to cope with the situation. Sometimes, men have worked collectively to meet the problems of a strange land. More often they have striven individually against greater hazards and difficulties than faced the eastern colonist. Nor have the settlers' roots been thrust particularly deeply into the land for which their grandparents rushed. Rootlessness is a feature of the second and third generations. Too many immigrants came to a land the characteristics of which they did not understand, and too many of their offspring have left before they have had time to absorb the new ideas and practices which are transforming agriculture.

Understanding comes with experience, aided by education and experiment. The wheat of the Prairies is still Canada's second main export by value. Stock and diversified crops can strengthen it but will not replace it. Industry may provide a valuable supplement to the Prairie economy, but the farm will prevail over the factory. Mines may be opened up, but their exploitation is a wasting process, while farming is a renewing one. Upon the basis of their combined agricultural and industrial potential, many have forecast for the Prairie cities a future in which they may outstrip those of the metropolitan east. But the cities of the east have the advantage of establishments and of markets. Possibilities and potentialities persist in the Prairies, but they are liable to be as deceptive as their distances. If there is one area in Canada which proves how well the best laid schemes of mice and men can go astray, it is the Prairie Provinces. They make their impact because of their difference from the rest of the federation. To paraphrase the words of Lord Bryce about Western America, 'The West may be called the most distinctly Canadian part of Canada, because the points at which it differs from the east are the points in which Canada as a whole differs from Europe'.

Detailed descriptions will be found in

D. F. Putnam (ed.), *Canadian Regions* (London, 1952)—Chapters 16, 17 and 18

while the historical background is dealt with agriculturally in

H. G. L. Strange, *A Short History of Prairie Agriculture* (Winnipeg, 1954)

and on a broader canvas by

W. A. Mackintosh and W. L. G. Joerg (ed.), *Canadian Frontiers of Settlement* (8 vols., Toronto, 1935)

Climatic problems are reviewed in

M. Sanderson, 'Drought in the Canadian North-West', *G.R.* (1948), 289–99
J. K. Villmore, 'The Nature and Origin of the Canadian Dry Belt', *A.A.A.G.* 46 (1956), 211–32

supplemented by

C. C. Spence, 'Water and Irrigation in Canada', *International Journal of Agrarian Affairs*, III, I (1961)

A fascinating compilation of topographical source material is

John Warkentin, *The Western Interior of Canada* (Toronto, 1964)

One aspect of the changing economy is considered in

A. V. Johnson, 'Relative Decline of Wheat in the Prairie Provinces of Canada', *E.G.* 24 (1948), 209–16

and many features of Manitoba are summarized in the

Economic Atlas of Manitoba (1960)

The *Association of Agriculture* maintains contact with a farm near Portage la Prairie, Manitoba

The North American Continent

14

The North American Continent—Problems in Stature and Status

The Middle Power and the Great Power—The Return to the Sea—
A Silhouette of a Land

The North American world is increasingly afflicted with problems of changing status. They are found at three levels. First, there are those which spring from the revaluation of resources and cause shifts in regional assessments within the political constituents of the sub-continent. Secondly, there are those which result from changing relationships between the political constituents. Thirdly, there are the problems born of the international shifts in power which affect the status of North America in a world context. All of these changes make for primary instabilities. Those at the world level cause especial stress to people who have hitherto found a measure of their stability in geographical isolation.

Changes in status are the result of many similar causes, but they have a different effect north and south of the 49th parallel. The causes are rooted in the nature of the land, the numbers of the people and the different rates of development of the two countries. They are frequently summed up in the differences between the great power and the 'middle power': the great power which must continuously review the world context and look to lands beyond the adjacent seas, and the middle power which is more influenced by than influencing the course of world events.

THE MIDDLE POWER AND THE GREAT POWER

As a political unit, Canada is a middle power: the U.S.A., a great power. Canada is more extensive in area than the U.S.A., but it bulks large in higher, less habitable latitudes and cannot therefore capitalize upon its space. It has but a tenth of the population of the U.S.A.: the number of its inhabitants approximately equals those of the Netherlands and Belgium or of the five Scandinavian countries collectively. Balancing area and population, Canada cannot be dismissed as a small power, but it cannot be described as a great one.

In its evolution, Canada holds an intermediary position: 'deadset in adolescence', as the poet Earl Birney put it in his incisive *Case History* of Canada. Dominion status came nearly a century after Republican independence, but it neither created nationality nor eliminated ethnographic dichotomy. Entrenchment has proved a stronger weapon for French-speaking Canada than secession, so that Canada has never had to struggle with secession as has the U.S.A. Two nations no longer war 'in the bosom of a single state' and Canadian nationality, essentially a twentieth-century product, is not without appeal to the Quebecois who is stronger in his resistance to American control than is the English-speaking Canadian.

The time lag in evolution by comparison with the U.S.A. is evident in a number of fields. For example, the tide of emigration to Canada gathered momentum more slowly; immigration restrictions have never been so stringent as in the U.S.A. and large numbers continue to be admitted. Toronto, the apotheosis of Anglo-Saxondom, is said to harbour a population of 500,000 New Canadians. For the U.S.A., with increasingly more American-born Americans, the assimilation of immigrants is of diminishing consequence: for Canada, it is still a matter of concern.

There has been a time lag in Canadian economic evolution. Quantitatively and qualitatively, many of Canada's resources await assessment, for they are located far beyond the margins of exploitation. The exploratory stage prevails over much of the country and the Department of Mines and Surveys co-ordinates in Ottawa the grand processional of technical exploration. The U.S.A. in its New Deal programmes preceded Canada through the phase of exploitation into that of conservation. Extensive areas of Canada constitute a *raubwirtschaft*, but the lessons of conservation are being widely learned. Canada remained a producer of primary materials for two generations after the U.S.A. had become a world industrial power. It has a world status as a reservoir of raw materials, but during the last generation has added manufacturing industry to primary production. The unprocessed products of forest, mine and farm still constitute the bulk of its exports, and as a trading country Canada is very sensitive to changes in demand for them. Sensitivity expresses itself in surplus-disposal difficulties, in support programmes for various sectors of the economy, and in cyclical unemployment exaggerated by the recurrent seasonal unemployment. Although it aspires to increase the range and variety of its exports through industrial processing, Canada is confronted with the problem of size in manufacturing activity. Scale of operations and size of home market are generally small compared with those of the U.S.A. The difficulties are exaggerated because the great power and the middle power are natural economic complements.

Finally, there has been a time lag in the politico-geographical evolution of Canada. Canada's isolation has been transformed with the opening of trans-polar air-routes. The strategic frontier has been shifted to its northern margins and Canada has become a buffer between the U.S.A. and the U.S.S.R. Alistair Cooke has spoken of Canada as a lifeline knitted thin but firm across the last vast band of plain and trees before the continent swings over to the Ice Age. A thousand miles north of the lifeline and in the Canadian Arctic is the ultimate strategic frontier of the twin nations of the continent.

This acquired location of military intermediacy gives new meaning to the inherited position of intermediacy between the U.S.A. and the United Kingdom. For Canada's final difficulty of adjustment is that it is intermediary between these two powers. There is no country with which it has closer contact than the U.S.A. and, reciprocally, there is no country with which the U.S.A. has more intimate relations than Canada. At the same time, Canada is the senior overseas member of the Commonwealth. Its two greatest trading partners and its two main sources of capital have been and are the U.S.A. and Great Britain; its principal source of population has been the British Isles and its chief route of emigration the U.S.A. Canada is intermediary in an eternal North Atlantic triangle. Its situation is more difficult than that of Mexico, because it has to contend with two disparate outside forces. It must continually resolve and respond to the differences of the U.S.A. and U.K. In the international scene, it may yet be forced to follow a Swedish way of neutrality.

THE RETURN TO THE SEA

In the U.S.A., at all levels, problems of changing status are shot through with a seaboard theme. When Benjamin Latrobe wrote to the *Transactions of the American Philosophical Society* in 1799 of the 'Neptunian origin of the whole country', his words had overtones beyond those of geological evolution. At a time when the frontier theme was claiming the growing attention of the country, Arnold Guyot still argued the superiority of the seaboard over all other zones. A tug-of-war between the interior and the coasts remains. Thus, despite the shift in the industrial centre of gravity in the north-east, the coast retains absolute domestic precedence. This is evidenced in the concentration of 30,000,000 people in the megapolitan area from Portland (Maine) to Richmond (Virginia), in the overwhelming commercial strength of New York City and the entrenchment of administrative power in Washington. The return to the seaboard is given new meaning with the phenomenal development of the maritime states of California and Texas. They represent powerful attractions to the coastlands of the Pacific and Gulf of Mexico. The revaluation of North America's 'fourth coast' of the Great Lakes through the St. Lawrence Seaway repeats the maritime theme in a different manner.

Once the U.S.A. acquired territories beyond its mainland rim, it forfeited a measure of its splendid isolation. The Spanish-American War of 1898 resulted in the acquisition of Caribbean and Pacific island possessions. Puerto Rico (1899) and the Philippines (1898–1946, when it became the Republic of the Philippines) were the foundations of an overseas 'empire' to which other constituents were added by purchase (e.g. the Virgin Islands from Denmark in 1917) and by negotiation. The Hawaiian archipelago (pop. c. 500,000) was formally annexed in 1898, together with a miscellany of Pacific islands, from Midway to Wake and Guam in the Caroline Islands and American Samoa (1899). The acquisition of overseas territories paralleled the emergence of the U.S.A. as an economic force of world stature, its transformation from a country of capital import to capital export, and its adoption of a policy of benevolent economic suzerainty, e.g. to islands such as Cuba.

The non-contiguity of these territories and the rise of North American markets overseas called for improved means of seaborne communication. In particular the U.S.A. became aware of the strategic inconveniences resulting from the isthmian separation of the Atlantic and Pacific Oceans. Negotiations to acquire the rights of Ferdinand de Lesseps eventually resulted in the American purchase of the Panama Canal Zone (1904). The 50-mile long canal was opened in 1914. By an agreement in 1930, Panama was divided between the Republic and a 10-mile-wide American Zone.

The return to the sea is in keeping with the tradition that a nation with the status of a world power must also command sea power. America's mercantile marine of 34,000,000 tons gives to it world leadership and its shipbuilding capacity is second to none. In naval strength, exploit and experiment, it asserts itself longitudinally as well as latitudinally. The precedent of the Monroe Doctrine directs its continuing attention to the seas about Latin America, while Yankee successors have not been unmindful that it was Nathan Palmer of Connecticut who first set foot upon the continent of Antarctica. America's massive expeditions to the Great White South demonstrate a persistent interest in the inheritance of the hemisphere's southern extremities. Simultaneously, the strategic revaluation of the hyperborean extremities through conquest of the air has restored an interest in seas and archipelagos made famous by the sagas of Frobisher, Franklin and Hudson. The Canadian Navy has forced the North-West Passage, but only as an exploit in calculated defiance of nature. Regular high-latitude movement is not upon the frozen seas; but above and soon, perhaps, below them.

America's maritime interests have reached their most extreme expression since the second world war, with the extension of bases to the seaboards of the Old World. In Europe, American bases are established through the N.A.T.O. organization; in the Far East, military aid varies in its scale of assistance and degree of control from Taiwan in the south to Korea in the north. It is a reality which accords almost if not quite with the prophecy of Timothy Dwight in his Goldsmithian poem *Greenfield Hill* (1794):

> Soon shall thy sons across the mainland roam
> And claim, on far Pacific shores, their home . . .
> O'er morn's pellucid main expand their sails,
> And the starr'd ensign court Korean gales.

In all its facets, the return to the sea of the U.S.A. could not fail to command the respect of the most distinguished maritime strategist in its history. And Jules Verne, not less than Admiral Mahony, might well marvel at the sub-polar exploits of the latter-day *Nautilus*.

A SILHOUETTE OF A LAND

The central chapters of this book have focussed upon the problems which continue to differentiate the ten principal regional units of the U.S.A. and Canada. To the extent that the problems persist, so do the old regional moulds into which the continent was cast historically. Yet the political unity of the continental area, the material wealth of its society and the

Z

intensity and spread of its means of communication all make for disintegration of the traditional regional frames. The essential physical geography prevails, names on the land evoke past distinctions, but the human scene is stamped with increasing uniformity. All sorts and conditions of men in all sorts and conditions of places prefer identical designs for living. As a result the human landscape is invested with a welcoming familiarity no matter how the arrangement of natural features varies. This has the effect of throwing the natural elements of the landscape into stronger relief.

The primeval and the sophisticated occur side by side with unexpected frequency in North America. Many people have wild country on their doorsteps. Cushioned against it more substantially than most of mankind, they are correspondingly more given to escaping into it. In the settled countryside, birds such as the naturalist J. J. Audubon drew with life-size fidelity a century ago may still delight and surprise; while creatures against whom pioneers protected themselves now enjoy the sanctuary of scores of impressive national parks and nature reserves. Within relatively easy access, there is enough wilderness to meet everyone's needs, so that hunting (called shooting in Europe) and fishing are for the many—not the few. And, if he wishes, the modern North American can still play the pioneer, since Canada and Alaska retain open frontiers. They have land where men have yet to tread and dangers no less real than those which confronted the Mackenzies, Selkirks and other explorers who are commemorated in river and range. Though most may be drawn towards what Edward Ullman calls the 'frontier of comfort', the scarcely occupied and barely named northlands remain territories of escape for latter-day Thoreaus from economic and social pressures of a more intense kind than those of the Old World.

North American man rejoices in his capacity to triumph over nature—to 'denaturalize' the land, as Robert Graves puts it. His belief in his ability to control the land is as strong as that of Soviet man. It is not surprising that much of the countryside has the appearance of being mastered rather than tamed. Yet, in the mastery of the land and in the conformity of its people there is found a measure of security. The search for security emerges in an essay by Thomas Wolfe entitled *God's Lonely Men*. It is his belief that most North Americans have yet to feel themselves at one with their land in the Old World sense. Only time can create this feeling. Small wonder that the American and the Canadian esteem the future over the past.

In this land of many characters, geographical breadth means more than historical depth. In North America, much change has been compressed into a very short time. The U.S.A. and Canada have achieved nationhood by revolution and evolution respectively: their peoples have compromised in federalism rather than yielding to sectionalism, their lands know conservation in place of exploitation. Production in the present surpasses the needs of the world's most affluent society, resources for the future are of an amplitude exceeded by no area of equal magnitude. The New World is still upon the threshold of its heritage and a silhouette of that which it might be. The true stature of North America remains to be revealed—for it is in the future that its peoples seek the image of themselves.

Appendix A

General Bibliographical Note

The most up-to-date contributions of the geography of North America are to be found in its journals. The principal of them (abbreviated in the text as noted here) are

> *The Geographical Review* (G.R.)
> *The Annals of the Association of American Geographers* (A.A.A.G.)
> *The Geographical Bulletin* (G.B.)
> *Economic Geography* (E.G.)
> *Canadian Geographical Journal* (C.G.J.)
> *Cahiers de Géographie de Québec*

There are many contributions by Canadians in

> *The Encyclopedia Canadiana* (Ottawa, 1958)

and as a supplement to any text on Canada, see

> *The Atlas of Canada* (Ottawa, 1957)

Still fundamental to American farming considerations is

> O. E. Baker, *Atlas of American Agriculture* (Washington, 1918–36)

Among recent publications or revisions of established texts dealing with North America are

> H. H. McCarty, *The Geographic Basis of American Economic Life* (New York, 1940)
> A. W. Currie, *Economic Geography of Canada* (Toronto, 1945)
> J. R. Smith, *North America* (New York, 1945)
> G. Taylor, *Canada* (London, 1947)
> C. L. White and E. J. Foscue, *Regions of Anglo-America* (New York, 1947)
> J. Gottman, *L'Amérique* (Paris, 1949)
> Ll. R. Jones and P. W. Bryan, *North America* (London, 1950)
> W. H. Parker, *Anglo-America* (London, 1962)
> N. J. G. Pounds, *North America* (London, 1955)

Earl B. Shaw, *Anglo-America, a Regional Geography* (New York, 1959)
J. H. Paterson, *North America: a Regional Geography* (Oxford, 1960)
H. Keenleysive and G. S. Brown, *Canada and the United States* (London, 1952)
J. Wreford Watson, *North America—its Countries and Regions* (London, 1963)

A most helpful approach to understanding the Canadian scene is

C. L. Blair and R. I. Simpson, *The Canadian Landscape. Map and Air Photo Interpretation* (Toronto, 1967)

A French Canadian view of Canada is provided by

L-E Hamelin, *Le Canada* (Paris, 1968)

There are four indispensable agricultural studies

L. Haystead and G. C. Fite, *Agricultural Regions of the U.S.* (New York, 1955)
F. Marschner, *Land Use and Its Patterns in the United States* (Washington, 1959)
A. N. Duckham, *American Agriculture, its Background and its Lessons* (London, 1959)
M. A. Tremblay and W. J. Anderson (ed.), *Rural Canada in Transition, Agr. Econ. Research Council of Canada*, 6, (Ottawa, 1966)
John Warkentin (ed.), *Canada: a geographical interpretation* prepared under the auspices of the Canadian Association of Geographers (Toronto, 1968)

Finally—a thoughtful book relevant to Chapter 14 and focused on conservation problems

Roderick Nash, *Wilderness and the American Mind* (Yale, 1968)

Appendix B

A Key to the Maps

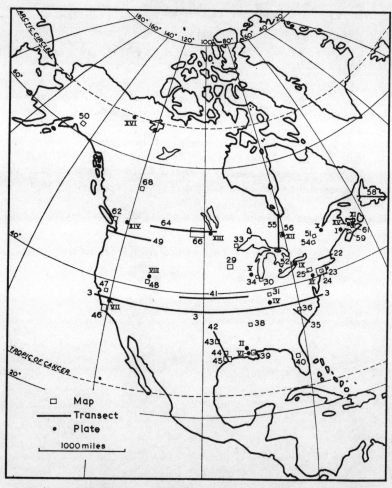

Figure 69 (*Appendix B*) *A key to the Location of Maps, Diagrams and Plates*

Appendix C

Principal Cities of U.S.A. and Canada according to Census Estimates

Census 1960
Standard Metropolitan Areas

NEW YORK	10,604,300
LOS ANGELES	6,683,563
CHICAGO	6,172,127
PHILADELPHIA	4,289,194
DETROIT	3,744,544
SAN FRANCISCO-OAKLAND	2,721,045
BOSTON	2,566,872
PITTSBURGH	2,395,249
ST. LOUIS	2,040,189
WASHINGTON	1,986,562
CLEVELAND	1,780,263
BALTIMORE	1,706,076
NEWARK	1,675,817
MINNEAPOLIS-ST. PAUL	1,477,080
BUFFALO	1,303,658
HOUSTON	1,234,868
MILWAUKEE	1,186,875
PATTERSON-CLIFTON-PASSAIC	1,180,186
SEATTLE	1,096,778
DALLAS	1,073,573
KANSAS CITY	1,027,562
SAN DIEGO	1,002,522

Estimate 1966
Census Metropolitan Areas

MONTREAL, PQ.	2,109,509
TORONTO, ONT.	1,824,481
VANCOUVER, B.C.	790,165
WINNIPEG, MAN.	475,989
OTTAWA, ONT.	429,750
HAMILTON, ONT.	395,189
QUEBEC, PQ.	357,568
EDMONTON, ALTA.	337,568
CALGARY, ALTA.	196,152
WINDSOR, ONT.	279,062
HALIFAX, N.S.	183,946
LONDON, ONT.	181,283
KITCHENER, ONT.	154,864
VICTORIA, B.C.	154,365
SUDBURY, ONT.	110,694
SAINT JOHN, N.B.	95,563
ST. JOHN'S, NFLD.	90,838

Index